Safety in the Lab

- Be sure to identify all safety equipment in the laboratory: eyewash station, fire extinguisher, first aid kit, broken glass container, and disinfectant solution.

- Do not eat, drink, or smoke in the laboratory.

- Always wear protective eyewear when appropriate. Many chemicals used in the laboratory can cause irritation to the eyes.

- Contact lenses should not be worn, wear eyeglasses instead.

- Do not apply cosmetics or take medications in the laboratory.

- Dress appropriately. Short pants and shoes with heels are not safe.

- Do not leave hot plates or open flames unattended.

- Make sure to always label test tubes, slides, and glassware properly.

- Long hair should be tied up so there is no danger of chemical contamination or catching fire.

- Always wear rubber or latex gloves when dissecting or working with caustic or toxic chemicals.

- Avoid contact with bodily fluids such as saliva, urine, and blood.

- Avoid wearing dangling jewelry.

- Alert the instructor to any medical condition that prevents you from performing any of the exercises.

- Immediately alert the instructor to any accidents, spills, broken glass, or injuries.

- Clean your work area with disinfectant after each lab period.

- Wash your hands with soap and water before leaving the laboratory.

In addition to these general laboratory guidelines, each exercise will alert you to specific precautions to take.

Human Biology

LABORATORY MANUAL

Charles J. Welsh

La Roche College, Pittsburgh, PA

JONES AND BARTLETT PUBLISHERS

Sudbury, Massachusetts

BOSTON   TORONTO   LONDON   SINGAPORE

World Headquarters

Jones and Bartlett Publishers
40 Tall Pine Drive
Sudbury, MA 01776
978-443-5000
info@jbpub.com
www.jbpub.com

Jones and Bartlett Publishers
Canada
6339 Ormindale Way
Mississauga, Ontario L5V 1J2
CANADA

Jones and Bartlett Publishers
International
Barb House, Barb Mews
London W6 7PA
UK

Jones and Bartlett's books and products are available through most bookstores and online booksellers. To contact Jones and Bartlett Publishers directly, call 800-832-0034, fax 978-443-8000, or visit our website www.jbpub.com.

Substantial discounts on bulk quantities of Jones and Bartlett's publications are available to corporations, professional associations, and other qualified organizations. For details and specific discount information, contact the special sales department at Jones and Bartlett via the above contact information or send an email to specialsales@jbpub.com.

Production Credits

Acquisition Editor, Science: Cathleen Sether
Managing Editor, Science: Dean W. DeChambeau
Editorial Assistant, Science: Molly Steinbach
Production Editor: Anne Spencer
Marketing Manager: Andrea DeFronzo
Text and Cover Design, Composition: Anne Spencer
Illustrations: Kate Ternullo, Imagineering, Elizabeth Morales, Graphic World
Photo Research Manager/Photographer: Kimberly Potvin
Photo Researcher: Christine McKeen
Cover Image: © Stephen Simpson, Getty Images, Inc.
Printing and Binding: Courier Stoughton
Cover Printing: Courier Stoughton

Library of Congress Cataloging-in-Publication Data
Welsh, Charles J.
  Human biology laboratory manual / Charles Welsh.-- 1st ed.
      p. cm.
ISBN-13: 978-0-7637-3843-3
ISBN-10: 0-7637-3843-3
  1. Human biology--Laboratory manuals. I. Title.
  QP44.W45 2006
  612.078--dc22
                                            2005036001
6048
Printed in the United States of America
10 09 08 07 06   10 9 8 7 6 5 4 3 2 1

Contents

Preface

This lab manual, to be used in Human Biology courses, was developed in conjunction with *Human Biology, 5th Edition*, by Daniel D. Chiras. It can also stand alone for use with other texts. I have successfully employed all of the exercises in the manual with students who had little or no background in the life sciences. The exercises are practical with regard to cost, set up, and implementation. Most are quite simple but a few do require some instructor preparation. Nearly all of the exercises can be completed in 1 1/2 to 2 hours and include easy-to-follow procedures for both instructors and students. For laboratory courses that are scheduled for 3 hours there is always the option of combining exercises.

The first lab combines the scientific method with the metric system. This is an integrative approach that is logical given that science strictly employs the metric system. While not unique, it is the better approach found in laboratory manuals. Combining microscopy with cell structure is a much more practical approach for this level because the ultimate goal is to gain experience identifying cells and tissues using a light microscope. This approach is rare in manuals, but in practice it seems to be preferable.

A feature of this manual is that most exercises are independent of each other and, after the first few introductory labs are completed, can be done in virtually any order. Moreover, several systems such as cardiovascular and digestive are expanded into two exercises, allowing for greater flexibility. Instructors can choose an integrative or separate approach. That is, structure and function can be studied together or separately.

Each exercise begins with learning objectives followed by a list of materials, a Safety Alert, an Introduction to review the concepts, and the Procedures section. Lab safety is emphasized in every exercise. Brief, boxed discussions titled "Of Interest" detail scientific and social aspects and "Clinical Considerations" describe medical issues. Graphs and space are provided throughout the manual so that students can record their responses and results. The manual is printed on perforated paper to allow for removal and submission to the instructor. The exercises end with Review Questions—10-15 questions to reinforce the principles covered in the lab.

Appendices at the end of the manual provide: 1) anatomical terminology, planes and directions, and 2) fetal pig dissection instructions, and 3) depictions of dominant and recessive genetic traits.

Charles Welsh

About the Author

Charles Welsh, LaRoche College, has been teaching Human Biology, Anatomy and Physiology, and Introductory Biology at the college level for 15 years. He has a BS and PhD in Biology. He is also a freelance technical writer and a Research Associate at the Carnegie Museum of Natural History in the Section of Birds where he studies Comparative Anatomy.

Acknowledgments

I would like to thank my wife, Lori, and my children Leeanna, Timothy, and Brady for their sacrifice during the preparation of the manuscript. This is as much their accomplishment as it is mine. I also would like to thank Dr. Robert Raikow, he saw the scientist/educator in me longer before most. His influence and wisdom abound throughout these pages. Scott Wells is thanked for supplying several crucial diagrams. A special thanks goes to the faculty in the Biology Department at La Roche College for their continued support. Pat Miller provided help during preparation of the manuscript, for which I am very grateful. William Daugherty is thanked for his encouragement and friendship.

Finally, I thank the entire editorial and production staff at Jones and Bartlett: Steve Weaver, Executive Editor for many supportive and insightful conversations, Dean DeChambeau, Managing Editor, for his ability to focus me, Cathleen Sether, Acquisition Editor for her support, and especially Anne Spencer, Production Editor, she has been a wonderful teacher and confidant throughout this project.

Charles J. Welsh

The Scientific Method and the Metric System

objectives

- To gain an understanding of and employ the scientific method
- To measure using the metric system
- To define and distinguish science, hypothesis, and theory

materials

- metric rulers and meter sticks
- goldfish
- graduated cylinders
- sugar cubes
- thermometers (in celsius)
- 1000-ml beakers
- hot plates
- distilled water
- ice bath
- scales and stopwatches
- coins and paperclips
- rubber or latex gloves

SAFETY ALERT!

You will be working with glassware (beakers). In the event that glass is broken, consult your instructor concerning cleanup procedures. The live specimens (goldfish) should be treated humanely and regarded as potentially infectious. Wear gloves when handling the fish. Exercise caution near the hot plates.

Introduction

Scientific Method

The term *science* comes from the Latin word *sciens,* which means to have knowledge. A modern definition of science could be this: a discipline that involves collecting and analyzing information (data) in an organized and logical way. This system is known as the "Scientific Method." It proceeds in several ordered steps for the purpose of drawing conclusions about the natural world (Figure 1-1).

A simple example will illustrate the process. The first step is to make an **observation.** You do this every day. For instance, you have probably observed that at least for a part of the year many trees have green leaves. In the second step, a **question** is asked about the observations made in step 1. In our example, you could ask this: Why are the leaves green?

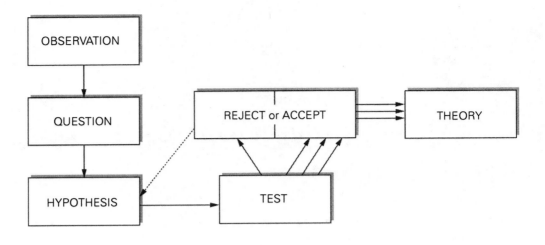

The third, and thought by many to be the most important step, is to formulate a **hypothesis**. The hypothesis is an attempt to answer the question in step 2 based on previous experience or data. Not only does the hypothesis have to be a reasonable answer, it also has to be testable. A solid hypothesis in this scenario would be this: The leaves are green because they contain chlorophyll, the pigment found in all other green plants.

The fourth step involves **testing** the hypothesis by **collecting more data**. This can be accomplished through experimentation or further observations. A simple chemical test can reveal the presence of chlorophyll.

Based on the results, you will draw a **conclusion**. That is, you will either **support or reject** your original hypothesis. This will not necessarily be the end of the inquiry. Scientific investigation is tentative. Therefore, the terms *proved* and *true* are rarely used because further data collection can potentially force you to alter or abandon your hypothesis.

Hypothesis to Theory

If many lines of evidence support the original hypothesis, it may be elevated to a **theory**. The theory of evolution provides a solid example. Many lines of evidence, including comparative anatomy, the fossil record, and DNA studies, support the contention that life on earth has evolved. We expand this discussion in Exercise 17.

The Metric System

Measurement is an important aspect of scientific investigation. Therefore, science strictly employs the **metric system** because it is all at once simple, logical, and consistent. First, it is a **decimal system**: it is based on multiples of 10. This makes conversions and calculations easy because it usually involves simply moving a decimal point to the right or left (**TABLE 1-1**).

In the metric system, the basic unit of **length** is the **meter**. Length is the most fundamental of all measurements. Not only is it used to measure one dimension such as height, it is needed to measure three-dimensional objects. These objects take up space and have a **volume**, measured in **liters**. You are probably familiar with buying beverages in 1- and 2-liter containers. Anything that has a volume will have a **mass (weight)**. The **gram** is the basic unit of mass. The metric system is built on this succession using **water** to standardize volume and mass. One **milliliter** (equivalent to one cubic centimter) of water has a mass of 1 gram (**Figure 1-2**).

TABLE 1-1

Basic Metric Units and Conversions

	Basic Unit	x1,000	1/100	1/1000	1/1,000,000
Length	Meter	Kilometer (km)	Centimeter (cm)	Millimeter (mm)	Micrometer
Volume	Liter	Kiloliter	Centiliter	Milliliter (ml) = (cc)	Microliter
Mass	Gram	Kilogram	Centigram	Milligram (mg)	Microgram

1ml H_2O = 1 cubic centimeter = 1gram

Figure 1-2 A cubic centimeter (cc) will hold a volume of water that will be 1 milliliter (ml) and have a mass of one gram.

Metric units are expressed in clear and consistent terminology. The standard U.S. system does not afford this. It is neither a decimal nor a consistent system. For instance, the term "ounce" is used to measure both volume and mass. **TABLE 1-2** compares the metric and standard U.S. systems and shows some common conversions.

TABLE 1-2 **Metric and US Equivalent Conversions.**

Metric		US Equivalent
1 Meter	=	39.3 Inches
2.54 cm	=	1 inch
1 Liter	=	1.06 Quarts
3.8 Liters	=	1 Gallon
1 Kilogram	=	2.2 Pounds
28.3 Grams	=	1 Ounce
Water freezes at 0° C	=	32° F
Water Boils at 100° C	=	212° F
Body Temperature is 37° C	=	98.6° F

Temperature and Time

Temperature is measured in degrees **centigrade** in the metric system. In fact, centigrade means "divided by 100." Again, water is used as a standard. In this system, water freezes at 0°C and boils at 100°C. Compare this with the freezing and boiling points of water using the Fahrenheit scale: 32°F and 212°F, respectively.

The basic unit of time is the **second**. Often in science, smaller units of time such as the millisecond, 1/1000 of a second, are measured.

procedure

Scientific Method

Each group will be given two goldfish, each in separate 1000-ml beakers filled with distilled water at room temperature.

1 Spend 5 minutes observing the appearance and behavior of the goldfish. While you do this, take caution not to disturb the goldfish. Here are some facts that will help you: Fish are "cold blooded," or poikilothermal. That is, they do not maintain a constant internal body temperature. Water holds more oxygen at lower temperatures. Fish extract their oxygen out of the water by running it over their gills. You can see this as a fish opens and closes the operculum (**Figure 1-3**). Fish tend to become more active as temperature increases.

Operculum

Figure 1-3

> **HINT**
> These are the things to look for: respiration rate (how often do you see its gills moving), whether the goldfish is moving slowly or quickly, whether it moves up and down or in circles.

2 Think about things that may affect the phenomena that you observed. Ask a question. For example, "What will happen if I . . . ? How will _____ affect . . . ?

> **HINT**
> Temperature, noise, light. (You will be supplied with ice and hot plates if you wish to study the effects of temperature change.)

3 Formulate a hypothesis to test. That is, based on your observations and questions about what may affect the goldfish behavior, come up with an answer.

procedure continued

4 Test your hypothesis.

5 Draw a conclusion: accept or reject.

Experimental Design

To test your hypothesis, you will need to design an experiment.

1 Make sure that you take good notes on what the fish does before the experiment (i.e., before you experimentally change something).

2 You must use a **control.** This is why you have been given two fish. The control fish will be the one to which you do nothing. It is used to make sure that something other than your experimental procedure does not have any effect. You are provided with two graphs to record data: one for the control and one for the experiemental fish.

Measuring Using the Metric System

Length

1 Using metric rulers and meter sticks, measure the length of objects in the lab: table, height of lab partners, width and length of room, etc. Record the results.

2 Uncoil a paperclip and measure its length. Measure the diameter of a few coins. Record the results. Measure (in cm) the height, width, and length of a sugar cube. **Note: multiply the height × width × length of the cube. This will give the cubic centimeters.**

Results:_____

Volume

Obtain a 100-ml graduated cylinder and fill it to the 50-ml mark. To do this, you will need to view the cylinder from the side to determine where the **meniscus** is (**Figure 1-4**).
You can now indirectly measure the volume of the sugar cube, coins, and paperclips using a method known as **displacement.** Drop the objects separately into the water. Then read the meniscus. Your reading minus the beginning 50 ml will give you the volume of the object. Record you results. **Note: Did the volume of the sugar cube in cubic centimeters equal its volume in ml?**

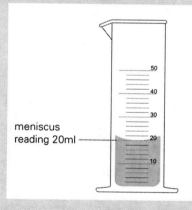

meniscus
reading 20ml

Figure 1-4

Results:_____

Mass

Using the scales, weigh the sugar cube, coins, and paperclips and any other objects that you have to the nearest 0.1 g. Record your results.

Results:_____

Is Medicine Science?

Although it employs knowledge and discoveries from science, is the practice of medicine considered science? The answer lies in the ability of medicine to conform to the scientific method. A hypothetical scenario will help decide whether medicine does conform.

Example: A health care professional at a college is alerted that many students on campus are experiencing a "red rash" over some of their skin. The professional is reasonably sure that he or she knows what it is and embarks on a treatment of administering antibiotics to 150 students. After 7 to 10 days, most of the students have no symptoms.

Science	Medicine
1. Observation.	1. Observation—red rash.
2. Question.	2. Question—what is causing the rash?
3. Hypothesis—answer to question.	3. Diagnosis—bacterial infection.
4. Testing—data collection with controls.	4. Treatment—give antibiotics.
5. Conclusion—accept or reject!	5. Conclusion—accept or reject diagnosis

You can see that medicine conforms to the scientific method very well for the first three steps. However, in step 4, science would dictate that control groups were used. That is, 50 patients would be given antibiotics, 50 would be given nothing, and 50 would be given a placebo to ward off the possible psychological effects of taking medication. Although this would constitute "good" science, it would be "bad" medicine as well as unethical to not treat all patients the same.

Name: _____   Lab Section: _____

⦚⦚⦚Review Questions

1. What is the second step in the scientific method?

   _____

   _____

2. How many millimeters (mm) are in 20 centimeters (cm)?

   _____

3. How much should one milliliter (1 ml) weigh?

   _____

4. How many meters are in one centimeter (1 cm)?

   _____

5. Based upon the discussion about temperature and water, which is hotter, 5° C or 32° F?

   _____

6. If your favorite NFL football team drafted a new linebacker who was said to be 6 feet 5 inches tall and weighed 250 pounds, you would have a fairly good idea of what he looked like. However, using your new knowledge of the metric system, give a rough conversion into meters and grams.

   _____

   _____

7. Why does science employ the metric system?

   _____

   _____

8. Do we really ever "prove" anything in science? Why?

   _____

   _____

9. Why must you use a control when designing an experiment?

   _____

   _____

Name: _____ Lab Section: _____

10. What was your hypothesis? How did you test it? Can you think of anything that may have caused experimental error?

_____

_____

_____

_____

_____

_____

_____

_____

_____

_____

_____

_____

_____

_____

_____

_____

_____

_____

_____

_____

Name: _____   Lab Section: _____

▕▏▎▍Exercise 1: Goldfish Experiment

Control Fish

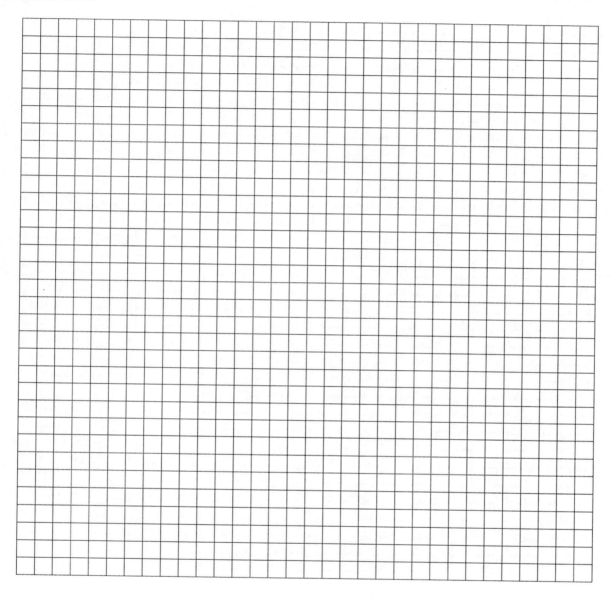

_____

Observations:

Name: _____   Lab Section: _____

‖‖‖‖Exercise 1: Goldfish Experiment

Experimental Fish

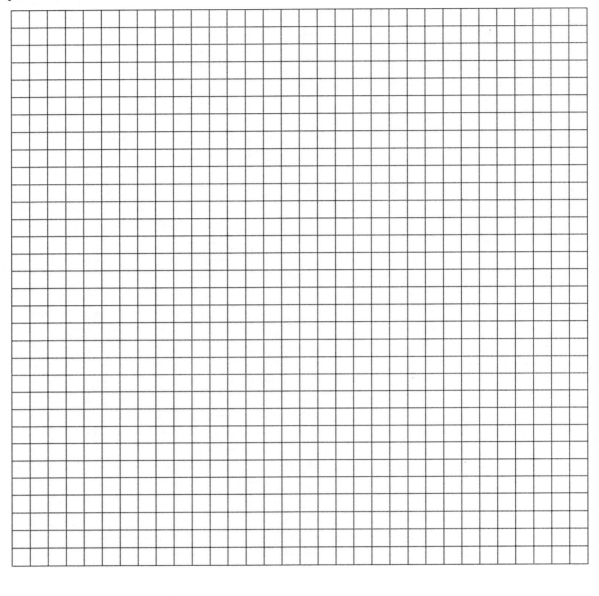

_____

Observations:

Chemistry of Cells: Macromolecules and pH

objectives

- To become familiar with the types of macromolecules found in cells
- To learn to perform a simple chemical test for macromolecules
- To employ the scientific method to discover macromolecules in common foods
- To define and measure pH

materials

- test tubes and racks
- Biuret reagent
- distilled water
- Benedict's solution
- brown paper bags
- iodine solution
- starch solution
- albumin solution
- glucose or fructose solution
- potatoes, onions, peanuts, bread, and apples
- knives or razor blades
- hot plates
- 500-ml beakers
- scissors
- pH paper
- baking soda
- milk
- vinegar

▼SAFETY ALERT!

You will be using glassware (beakers and test tubes). In the event that any glass is broken, consult your instructor concerning cleanup procedures. Exercise caution near the hot plates. Take care when using scissors, knives, and razor blades. In the event that you or your lab partners are cut, consult your instructor concerning first aid procedures.

⦀Introduction

All matter, including that found in cells, is composed of atoms and molecules. Living systems are set apart, however, because they are built from **macromolecules**. These large **carbon-based**, or **organic**, molecules are highly organized. Together they orchestrate the complex processes carried on inside of cells. They include **carbohydrates, lipids (fats),** and **proteins**. The chemistry of cells is also dependent upon maintaining proper **pH** (acid–base) levels. In this lab, you will perform simple chemical tests to reveal the presence of macromolecules and measure the pH of various common solutions.

Carbohydrates

Carbohydrates are molecules that you have probably heard of with regard to diet and nutrition because they are found in breads, grains, and potatoes. They are also known as **saccharides**, or **sugars**. They contain carbon, hydrogen, and oxygen in this ratio: $C:H_2:O$. The ratio can also be written as $C:H_2O$. That is, for every one carbon atom, there will be one molecule of water.

Glucose ($C_6H_{12}O_6$) is the standard carbohydrate by which all others will be named and classified. It is known as a **monosaccharide** (single sugar). Saccharides are classified based on how many glucose or other monosaccharide molecules are needed to build them. Fructose is also a monosaccharide that shares the formula $C_6H_{12}O_6$. Glucose and fructose are two of the preferred molecules that cells use to make energy for all metabolic processes.

Disaccharides (double sugars) are molecules built from two ("di") monosaccharides. Sucrose, common table sugar, is a disaccharide that is built from the joining of a fructose and a glucose molecule:

$$C_6H_{12}O_6 + C_6H_{12}O_6 \rightarrow C_{12}H_{12}O_{11} + H_2O$$

Figure 2-1 shows the structures of the molecules involved.

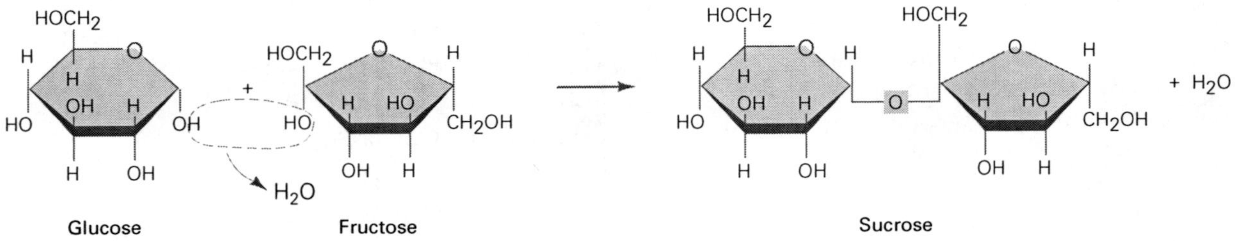

Figure 2-1 Formation of sucrose.

Sucrose has the formula $C_{12}H_{22}O_{11}$ and does not conform to the 1:2:1 ratio. However, if the entire system is taken into account, the water that has been removed would restore the ratio. It is necessary to lose a molecule of water in order to open available bonds for the joining of the two monosaccharides. This is known as a **dehydration synthesis**: losing water to build bigger molecules.

Starch is a **polysaccharide** (many sugars) built from hundreds of glucose molecules. Plants store glucose as starch. This is why we eat grains and potatoes. Humans and other animals store their glucose in the form of the polysaccharide **glycogen** (**Figure 2-2**).

Starch and glycogen can be broken down to release the glucose for energy production by way of a hydrolysis reaction:

$$starch + H_2O \rightarrow glucose$$

This is the opposite of a dehydration synthesis.

Proteins

Proteins are large molecules that play a variety of vital roles in cells. The majority of proteins serve as **enzymes**. These are catalysts that enhance the productivity and efficiency of all chemical reactions inside cells. In fact, without enzymes, reactions could never proceed fast enough to sustain living systems.

Lactase is an enzyme that aids in the chemical digestion of **lactose**, the sugar found in dairy products. The condition known as **lactose intolerance** is caused by a deficiency in the production of lactase.

Proteins are also used to build many of the supporting structures in cells and tissues. **Collagen** is a rigid protein found in skin, hair, nails, and bones. The proteins **actin** and **myosin** play a major role in muscle contraction.

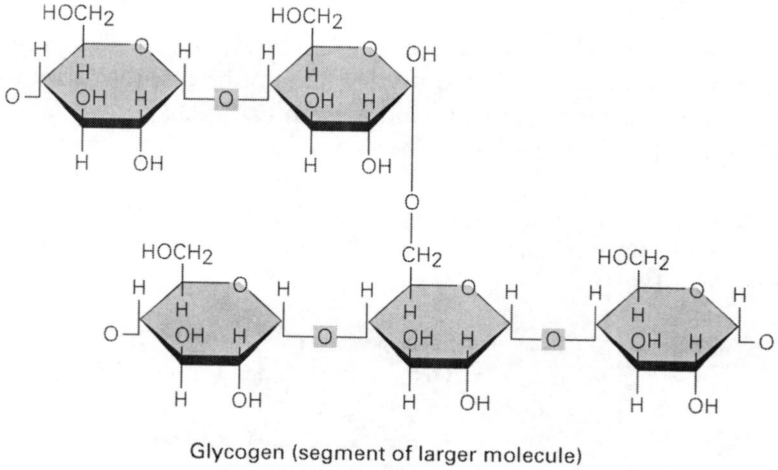

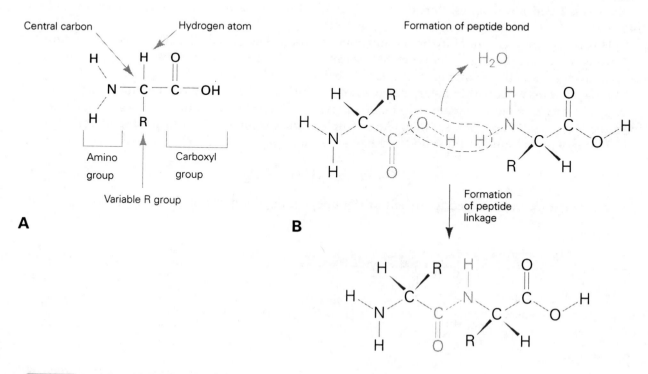

Figure 2-2 Polysaccharides

Some proteins, such as **insulin**, serve as messengers. Insulin is responsible for maintaining the proper levels of glucose in the blood stream. **Diabetes** is a disease in which insulin production is reduced or has altogether ceased.

Still other proteins will perform various functions in cell membranes. This is discussed in Exercises 3 and 4. Proteins are constructed from smaller building blocks known as **amino acids** (**Figure 2-3**).

Figure 2-3 A. Individual amino acid. B. Formation of the peptide bond.

Lipids

Lipids are commonly referred to as **fats**; however, they also include **oils**, such as vegetable oil, and **waxes**. They all share the unique property of **insolubility** (cannot be dissolved) in water. Like carbohydrates, they are composed of carbon, hydrogen, and oxygen (Figure 2-4).

Adipose tissue is comprised of fat molecules. The fat serves to insulate, cushion, and store energy.

Some of the more complex of the lipids are the **phospholipids**. They make up the bulk of cell membranes where they provide support and protection for cells. We will study membrane structure and function in Exercises 3 and 4.

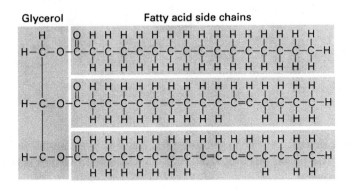

Figure 2-4   A lipid molecule.

pH

pH is the measure of acidity or alkalinity of a solution. Acids are compounds that release H^+ (hydrogen) ions in solution whereas bases release $-OH$ (hydroxide) ions. The pH scale ranges from 0 to 14 (Figure 2-5), with pH 7 being neutral. Compounds with a pH between 0 and 7 are acids. Those with a pH between 7 and 14 are bases. Pure deionized, distilled water has a pH of 7. It is neither an acid nor a base.

pH is actually a measure of hydrogen ion concentration. The greater the H^+ ion concentration, the stronger the acid. Living systems are very sensitive to changes in pH because hydrogen ion concentration can dramatically disrupt the chemical balance inside of cells.

Living systems exist within a very narrow range of pH, between 6 and 8. Therefore, they will employ chemicals called buffers to help maintain pH. Buffers accomplish this by donating or accepting either H^+ or $-OH$ ions in a solution. An example is the **carbonic acid/bicarbonate** buffering system. This system is found in the blood and helps maintain pH at about 7.4:

$$H_2CO_3 \longleftrightarrow H^+ + HCO_3^-$$

This reaction can be shifted to the left or right depending upon changes in pH.

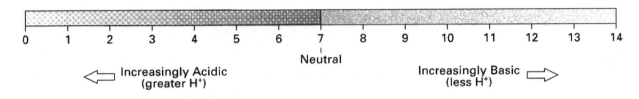

Figure 2-5   pH scale.

procedure

Test for simple sugars

1 In the presence of monosaccharides and some disaccharides, the Blue Benedict's solution will change color. Depending on the concentration of the sugar, the color change can range from green to red. Green indicates a low concentration, and a change to red reveals a high concentration.

    a. Fill two test tubes each with 3 ml of Benedict's reagent.

    b. To one tube, add 2-ml distilled water, and to the other, add 2-ml glucose solution.

    c. Heat both tubes in boiling water for 5 to 10 minutes.

    d. Record the results.

Test tube with Benedict's and water:_____

Test tube with Benedict's and glucose: _____

Test for starch

2 An iodine solution will turn from light brown to a blue, gray, or black color in the presence of starch.

    a. Fill two test tubes each with 3 ml of iodine solution.

    b. To one tube, add 2-ml distilled water, and to the other, add 2 ml of the starch solution. Make sure to label the tubes.

    c. Record the results.

Test tube with iodine and water: _____

Test tube with iodine and starch: _____

Test for proteins

3 Biuret reagent, which is blue, will turn pink or purple in the presence of amino acids and proteins, respectively.

    a. Fill two test tubes each with 3-ml Biuret reagent.

    b. To one tube, add 2-ml distilled water, and to other, add 2 ml of albumin solution. Be sure to label the tubes.

    c. Record results.

Test tube with Biuret reagent and water:_____

Test tube with Biuret reagent and albumin:_____

Test for lipids

4 Lipids will leave an "oily" smudge on brown paper.

   a. Obtain two small squares of brown paper.

   b. On one square, make a fingerprint using distilled water. On the other, make the same print with vegetable oil.

   c. Wait 5 minutes, and describe the results.

Results:_____

Test food products for carbohydrates, proteins, and lipids

5 You will be using several food products listed in Table 2-1 and any others your instructor provides for you. For each food, you will be testing for the presence of simple sugars, starches, lipids, and proteins. You will use the chemical tests that you performed in the first part of this exercise. You can employ the scientific method here.

   • **Observe** that different foods are eaten to obtain different macromolecules (nutrients).

   • **Ask** the following: Which foods contain which nutrients?

   • Formulate the **hypothesis**: I believe apples contain _____.

   • Test the hypothesis using the procedure for macromolecule detection.

   • **Accept** or **Reject!**

   a. Take a small (pea-sized) sample of each food product, and crush them separately into 15 ml of distilled water using a mortar and pestle.

   b. For every food product solution, pour 4 ml into three test tubes.

   c. Use one test tube for a simple sugar test, one for a starch test, and one for a protein test.

   d. Use the remainder of each solution to perform the lipid test on brown paper.

   e. Record results in **TABLE 2-1**.

TABLE 2-1	Results of Testing Food Products for Macromolecules			
	Simple Sugar	Starch	Protein	Lipids
Potato				
Apple				
Onion				
Peanut				
Bread				

6 Testing pH

   a. Pour about 200 ml of each of the solutions to be tested for pH in separate 500 ml beakers.

   b. Test the pH of each solution using pH paper and the corresponding charts it came with.

   c. Record the results in **TABLE 2.2**.

7 Buffers

a. Mix a tablespoon of baking soda into 200 ml of the distilled water you used to initially measure pH.

b. Measure the pH.

c. Record results in Table 2.2. If the pH changed, can baking soda be considered a buffer?

TABLE 2-2	Results of Testing pH of Solutions	
Solution	**pH**	
Distilled water		
Tap water		
Vinegar		
Milk		
Distilled water and baking soda		

CLINICAL CONSIDERATIONS

Diet, Nutrition, and Macromolecules

Vegetarian diets appeal to people for a variety of reasons. Some have ethical issues with the treatment and slaughter of agricultural animals, whereas others are concerned about ingesting too much fat and cholesterol. Still others may genuinely dislike meats. Vegetarians, however, must become informed about basic nutrition. That is, they must learn which food sources will ensure that they receive all of the necessary nutrients in their diets. For instance, grains, fruits, and vegetables are rich in carbohydrates (starch and sugar) but are low in proteins.

Proteins, which are actually ingested to acquire essential amino acids, can be found in nuts, milk, eggs, and meats. Some vegetarians eliminate all animal products from their diets, including milk and eggs. Therefore, they must develop diets that incorporate a mixture of plant products, such as corn and beans. Each of these contains complimentary essential amino acids that the other does not.

There are usually plenty of lipids in fruit and vegetables to sustain cellular needs. However, cholesterol, which is needed to build cell membranes, is found in only animal products. The liver can produce cholesterol if necessary, however, the occasional egg or glass of milk is recommended for those interested in a vegetarian lifestyle.

Name: _____   Lab Section: _____

|||||| Review Questions

1. What are the building blocks of proteins?

   _____

   _____

2. Which disaccharide is common table sugar?

   _____

   _____

3. Where are phospholipids found in cells?

   _____

   _____

4. What are chemical reactions called that "kick out" water to build bigger molecules?

   _____

   _____

5. The failure to produce the protein insulin leads to what disease?

   _____

   _____

6. What reagent would you use to test for the presence of proteins?

   _____

   _____

7. Which types of proteins speed up chemical reactions?

   _____

   _____

8. What is the chemical formula for fructose?

   _____

   _____

9. Should a glucose solution turn blue and black in the presence of iodine?

   _____

   _____

Name: _____   Lab Section: _____

10. What was the connection between using distilled water in the test tubes when testing for the presence of starches, sugars, and proteins and using distilled water to make solutions when testing food?

   _____

   _____

11. Why does water have to be "kicked out" to join two molecules?

   _____

   _____

Cell Biology and Microscopy

- To become familiar with the structure and function of cell membranes and organelles
- To learn the parts of the compound light microscope and how to use it
- To make wet mounts using animal and plant cells
- To study the stages of mitosis

- cell models
- microscopes
- toothpicks or cotton swabs
- slides and cover slips
- iodine solution
- methylene blue
- slides of letter *e*
- slides of overlapping colored threads
- slides of whitefish mitosis
- onions

SAFETY ALERT!

Handle the microscopes carefully, and always carry them with two hands. Slides are made of glass. Even when not broken, the edges are sharp. In the event that a slide is broken, inform the instructor and seek guidance concerning cleanup procedures. Exercise caution when using the toothpicks for the cheek cell exercise. Do not share toothpicks, as saliva and other bodily fluids can pose a health risk.

‖‖‖‖ Introduction

Cells are the smallest units that exhibit all of the characteristics associated with living systems: **growth, reproduction, metabolism (nutrient cycling and energy production)**, and **response to the environment**. Therefore, the study of cells continues to shed light on many basic questions concerning living systems. In this exercise, you will become familiar with the structure and function of cells through examination of diagrams, models, and microscopic specimens.

Cell Structure and Function

The insides of cells are filled with **cytoplasm**, or "cell fluid." It is made up of mostly water but also contains the type of macromolecules that were discussed in Exercise 2.

Organelles, "little organs," are structures also found in the cytoplasm that perform various vital functions. These are shown in **Figure 3-1**, and their functions are listed in **TABLE 3-1**.

Cells are surrounded by the **plasma membrane**. It provides protection and support and controls passage into and out of a cell. Membrane transport will be studied in Exercise 4.

TABLE 3-1	Overview of Cell Organelles	
Organelle	Structure	Function
Nucleus	Round or oval body; surrounded by nuclear envelope	Contains the genetic information necessary for control of cell structure and function; DNA contains hereditary information
Nucleolus	Round or oval body in the nucleus consisting of DNA and RNA	Produces ribosomal RNA
Endoplasmic reticulum	Network of membranous tubules in the cytoplasm of the cell. Smooth endoplasmic reticulum contains no ribosomes. Rough endoplasmic reticulum is studded with ribosomes	Smooth endoplasmic reticulum (SER) is involved in the production of phospholipids and has many different functions in different cells; round endoplasmic reticulum (RER) is the site of the synthesis of lysosomal enzymes and proteins for extracellular use
Ribosomes	Small particles found in the cytoplasm; made of RNA and protein	Aid in the production of proteins on the RER and polysomes
Polysome	Molecule of mRNA bound to ribosomes	Site of protein synthesis
Golgi complex	Series of flattened sacs usually located near the nucleus	Sorts, chemically modifies, and packages proteins produced on the RER
Secretory vesicles	Membrane-bound vesicles containing proteins produced by the RER and repackaged by the Golgi complex; contain protein hormones or enzymes	Store protein hormones or enzymes in the cytoplasm awaiting a signal for release
Food vacuole	Membrane-bound vesicle containing material engulfed by the cell	Stores ingested material and combines with lysosome
Lysosome	Round, membrane-bound structure containing digestive enzymes	Combines with food vacuoles and digests materials engulfed by cells
Mitochondria	Round, oval, or elongated structures with a double membrane. The inner membrane is thrown into folds	Complete the breakdown of glucose, cellular respiration
Cytoskeleton	Network of microtubules and microfilaments in the cell	Gives the cell internal support, helps transport molecules and some organelles inside the cell, and binds to enzymes of metabolic pathways
Cilia	Small projections of the cell membrane containing microtubules; found on a limited number of cells	Propel materials along the surface of certain cells
Flagella	Large projections of the cell membrane containing microtubules; found in humans only on sperm cells	Provide motive force for sperm cells
Centrioles	Small cylindrical bodies composed of microtubules arranged in nine sets of triplets; found in animal cells, not plants	Help organize spindle apparatus necessary for cell division

The plasma membrane is built from two layers of phospholipids molecules. This is often referred to as the "**phospholipid bilayer membrane.**" Found embedded within the membrane are proteins that act as channels, receptors, markers, and enzymes (**Figure 3-2**). This has led to the development of the "**fluid mosaic model**" theory of the structure of the cell membrane. It means that the membrane is "**flexible**" and "**not uniform**" throughout.

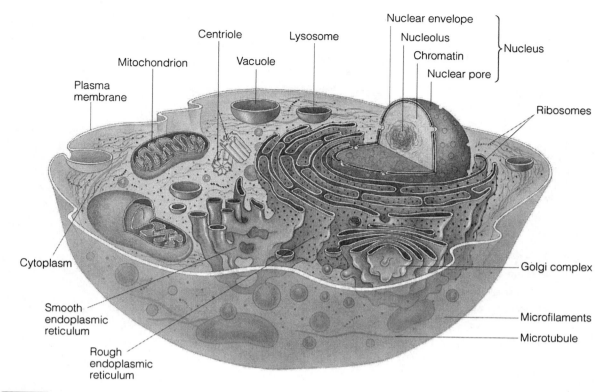

Figure 3-1 The structure of the cell.

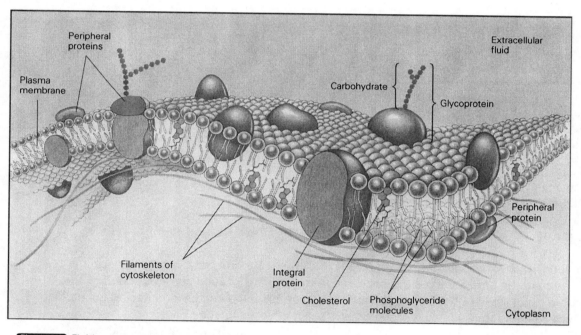

Figure 3-2 Fluid mosaic model of the plasma membrane of animal cells.

Microscopy

To view most cells, you have to use a microscope. In fact, it was the invention of the microscope that led to the development of the **Cell Theory**: (1) All organisms are built from cells. (2) The cell is the smallest, most basic unit of life, and (3) only cells can give rise to other cells. In this exercise, you will study the structure of cells using the compound light microscope (**Figure 3-3**). As the name implies, it employs a light source and at least two lenses to illuminate and magnify specimens. **TABLE 3-2** lists the parts and functions of a typical compound light microscope.

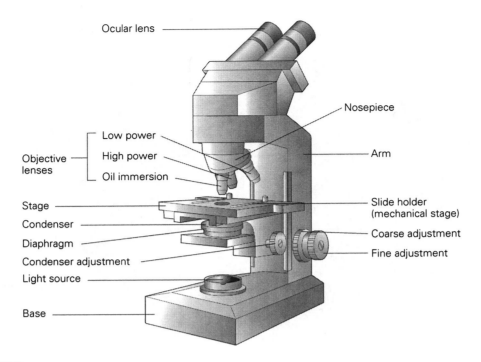

Ocular lens

Nosepiece

Low power
High power
Oil immersion

Objective
lenses

Arm

Stage
Condenser
Diaphragm
Condenser adjustment
Light source

Slide holder
(mechanical stage)

Coarse adjustment

Fine adjustment

Base

Figure 3-3 Compound light microscope.

procedure

Cell structure and function

1 Using any available models and diagrams in your textbook and Figure 3-1, identify all of the structures and their functions listed in Table 3-1.

Microscopy

2 Use both hands when carrying a microscope. Firmly secure one hand under the base and the other around the arm. Examine the microscope to learn the location and function of all the essential parts listed in **TABLE 3-2**. At your lab station, plug the microscope into an outlet. You are now ready to turn the microscope on and begin developing you skills.

Magnification

As mentioned in the introduction, a compound microscope employs at least two lenses. The first lens will be the **ocular**, or **eyepiece**. The ocular lens usually has a magnification power of 10. This is written as 10×. The other lenses are objective lenses. They will have magnification powers of varying degrees. They usually will range from 4× to 100×. The 4× will be the low power, or **scanning lens**. The higher power lenses will usually be 10×, 40×, and 100×. In this exercise, you will not use the 100× power lens often referred to as an oil immersion lens. It is for viewing very small cells such as bacteria.

TABLE 3-2	Microscope Parts and Their Functions

Parts	Function
1. Ocular, or eyepiece	A lens of a given magnification, which is probably engraved on the rim-for example, 10×. (You may also see a pointer embedded in the ocular. This is used to aid you in indicating specific locations within the microscopic field.)
2. Revolving nosepiece	This plate, which is capable of rotation, allows you to utilize objectives of different magnifications.
3. Objectives	Lenses of varying magnifications. The values (such as 4×, 10×, 43×, and 100×) are usually engraved on the objectives.
Scanning	Lens with the least magnification (often 4×).
Low-power	Lens with greater magnification than scanning objective (often 10×).
High-power	Lens with greater magnification than high power objective (often 43×).
Oil-immersion	Lens with greatest magnification (often 100×). You will infrequently use this lens in this course. Proper techniques must be demonstrated by your instructor prior to its use.
4. Arm	Handle for holding and positioning the microscope.
5. Stage	Platform on which the slide is positioned for focusing. Note that your scope may have clips for anchoring the slide or may have a mechanical slide holder. Obtain a blank slide and fit it into the holder. The slide can be moved forward, backward, and from side to side by rotating the small knobs on the underside or on top of the stage. If your microscope is not equipped with a mechanical stage, you will need to move the slide manually.
6. Coarse adjustment knobs	Large knobs on both sides of the base of the arm that allow for initial focusing of the object to be viewed.
7. Fine adjustment knobs	Small knobs on both sides of the base of the arm that allow for refinement of detail in focusing.
8. Condenser	A lens system that concentrates light from the illumination source so that a cone of light fills the aperture of the objective. After checking to see that the light bulb is on, move the condenser up and down and note the varying intensities of light visible through the ocular. Your microscope may not be equipped with a movable condenser.
9. Diaphragm	A plate with an aperture allowing for varying amounts of light to pass through the specimen. Open and close the diaphragm by adjusting it with its handle so that varying intensities of light are visible through the ocular.
10. Base with illuminator	Platform on which the microscope is structured, usually containing an electric light source.

The **total magnification** of the specimen will be the power of the ocular lens multiplied by the power of the objective lens. For example, if the ocular is 10× and the objective lens in use is 10×, the total magnification of the specimen will be 10× times 10×, or 100×. This can be written as a simple mathematical equation:

$$10X \times 10X = 100X$$

Examination of Prepared Slides

You will begin by examining a few prepared slides. The first slide to examine will be that of the letter *e*. As simple as this may seem, this slide will demonstrate one of the fundamentals of microscopy: the image you view will be **inverted** and **backward**.

1   Before you put the slide of the letter *e* on the microscope, hold it up to the light to view the image. Does it look like the letter as you would write it?

2   Turn the microscope on. The light should now be on.

3   Turn the course focus so that the stage and objectives lenses are as far apart as possible.

4   Put the slide on the stage.

5   Carefully click the lowest power (scanning) lens into place.

6   While looking into the eyepiece, slowly turn the course focus until a rough image comes into view. This may take some time as you become accustomed to using the lenses. You probably also need to move the stage to the left and right and back and forth to center the image.

7   After you have a rough image, use the fine focus to make the image as clear as possible. You may need to use the diaphragm to adjust the level of light.

Using Higher Powered Lenses

Switching to an objective lens with a **higher magnification** will demonstrate another fundamental characteristic of compound light microscopes: They are **parfocal**. That is, if an image is in focus under low power and you switch to a higher power, the image should still roughly be in focus. You should only have to make small adjustments with the fine focus to view a clear image.

1   Switch to the next higher magnification. Be careful to avoid "smashing" the objective lens into the slide. The higher the magnification, the longer the lens!

2   Using the fine focus bring the letter *e* into clear focus.
    Did you have to adjust the level of light as you increased the magnification?

Microscopic Specimens Are Three Dimensional

Even objects small enough to fit onto a slide and be viewed under a lens will have a three-dimensional structure. That is, they will have a **length, width,** and **depth**. This can easily be demonstrated by viewing a slide with three overlapping colored threads.

1   Place a slide with the three colored threads onto the stage and view under low power.

2   Switch to one of the higher power lenses. Slowly rotate the fine focus. As one thread comes

into clear view, the others should appear "out of focus." What does this show you about microscopic specimens?

3 Try to determine the order in which the threads were put onto the slide. Which is on the bottom, the middle, and the top?

Preparation of a Wet Mount

You can easily prepare a specimen for microscopic examination. In this procedure, you will actually have the opportunity to view cells from your own body. These cells will come from the lining of the inside of your cheek.

1 Obtain a microscope slide, a cover slip, and a sterile toothpick or cotton swab.

2 Using the toothpick or swab, gently scrape the inside of your cheek.

3 Smear the toothpick or swab onto the slide.

4 Add a drop of methylene blue or iodine solution to the specimen.

5 Carefully lower a cover slip on the slide at a 45-degree angle in order to minimize air bubbles (**Figure 3-4**).

Figure 3-4 Preparation of a wet mount.

6 Put the slide onto the microscope, and view it under low power. What do you see?

7 Switch to higher magnification. You should be able to view darkly stained ovals in the middle of the cells. These are the nuclei.

Further Support for the Cell Theory

The cells in your body are typical animal cells. However, according to the cell theory, all organisms are built from cells. In order to demonstrate this, you will make a wet mount with an onion. This will show typical "plant" cells.

1 Slice off as thin of a piece of onion skin as you can using a small knife or razor blade.

2 Place the piece of onion on a slide and add a drop of methylene blue or iodine solution.

3 Place a cover slip over the onion piece as you did in the cheek cell procedure.

4 Observe the specimen under the microscope using low and high power.

5 Again, you should see a large nucleus in each cell.

Mitosis

A cell spends most of its life engaging in metabolic activities such as building proteins and storing energy. At some point in its life, it engages in the process of mitosis, or cell division. It is the process by which new cells arise. It is both reproduction at the cellular level and how body tissues grow larger. Observe **Figure 3-5** and the slides of whitefish mitosis to identify the following stages of mitosis.

1 **Interphase.** This is actually the nondividing phase of the cell. However, during this time the DNA, which is seen as a dark mass called chromatin, replicates.

2 **Prophase.** The nucleus begins to disappear and the DNA begins to organize as chromosomes.

3 **Metaphase.** The chromosomes line up on the equator of the cell.

4 **Anaphase.** The chromosomes begin to pull apart. Half will go to each pole (side) of the cell.

5 **Telophase.** The two new nuclei begin to reappear, and the chromosomes disappear as the DNA is again confined to the nuclei. Eventually the two new cells will split apart.

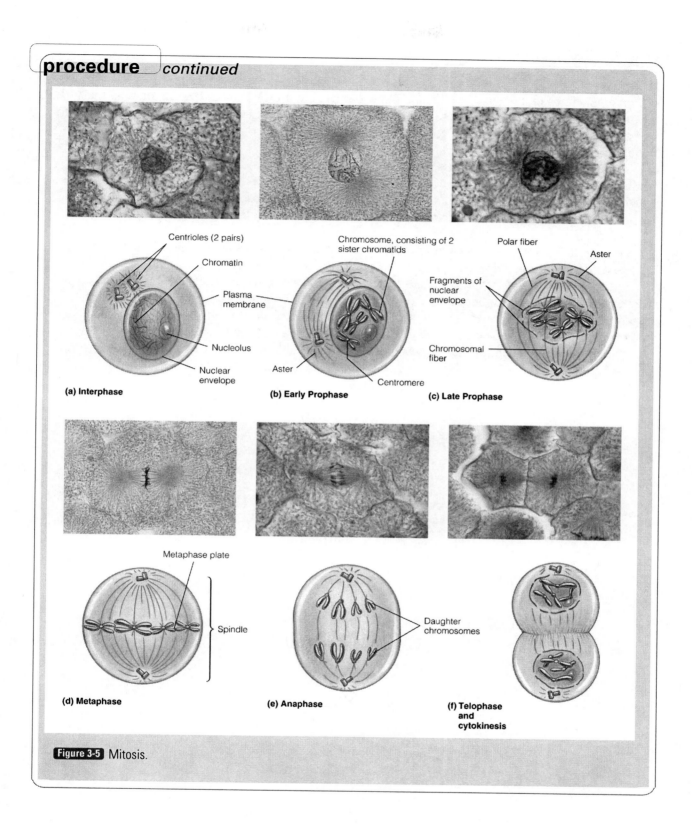

Figure 3-5 Mitosis.

Cancer

Cancer is a term that is used to describe many diseases that are characterized by an abnormally high rate of cell division, or mitosis. This usually will result in the development of a tumor, a growth of odd tissue. A **benign** tumor is not cancerous and does not grow rapidly. Cancerous tumors are **malignant** and grow and spread rapidly.

Cancerous tumors are themselves usually harmless. That is, they do not release toxins or other chemicals that disrupt metabolism. However, they become problematic when they compete with normal tissue for nutrients and space. This may result in the death of normal tissue and the beginning of pain and other physiologic problems. Tumors can also cause extreme pain if they press on nerves.

Causes of cancer are many and varied. Environmental factors usually involve chemical agents know as **carcinogens**. They can cause permanent changes in the DNA in cells that can lead to abnormal growth. Other carcinogens are thought to activate normally "quiet" genes. When activated, these are known as **oncogenes** and can cause a cell to undergo rapid, uncontrolled division. Carcinogens include cigarette smoke and ultraviolet radiation from the sun and many organic chemicals such as acetone (one of the chemicals found in nail polish remover).

Treatments for cancer include surgical removal of tumors, chemotherapy, and radiation therapy.

Name: _____   Lab Section: _____

‖‖‖ Review Questions

1. What two types of macromolecules are found in the plasma membrane?

   _____

   _____

2. If you wanted to view a specimen magnified by 400 times and the eyepiece has a magnification of 10×, which objective lens would you use?

   _____

   _____

3. What is the only organelle you saw in the wet mount of the cheek cells?

   _____

   _____

4. Which organelle is the site of cellular respiration?

   _____

   _____

5. Examination of the thread under the microscope was an exercise to illustrate what?

   _____

   _____

6. Where in a cell is DNA found?

   _____

   _____

7. In which stage of mitosis do chromosomes line up on the equator of the cell?

   _____

   _____

8. What does "fluid mosaic" mean?

   _____

   _____

Name: _____ Lab Section: _____

9. Why was it important to view an animal and a plant cell?

_____

_____

10. List the three parts of the Cell Theory.

_____

_____

_____

11. Draw a diagram of cell that is in the metaphase stage of mitosis.

12. Label the structures on the cell diagram.

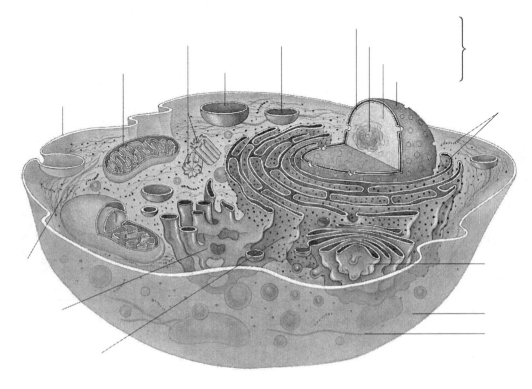

Membrane Transport: Diffusion and Osmosis

objectives

- To define diffusion and osmosis and give examples of each
- To gain understanding of selective permeability
- To define and distinguish isotonic, hypotonic, and hypertonic

materials

- chicken eggs with the shell dissolved away
- methylene blue
- scale
- fresh elodea leaves
- sucrose solutions: 0%, 10%, 20%, 30%, and 40%
- light microscopes
- 500-ml beakers
- 15% NaCl solution

⚠ SAFETY ALERT!

You will be working with glassware (beakers) and microscope slides. Even slides that are not broken are sharp. In the event that any glass is broken, seek guidance from your instructor concerning cleanup procedures.

▥ Introduction

As discussed in Exercise 3, **cells** are the smallest entities that exhibit all of the characteristics associated with living systems. The **selectively permeable cell membrane** controls the passage of ions and molecules into and out of cells. This will have a major effect on all of the characteristics of living systems. It is for this reason that studying the passage across cell membranes is of such interest.

Diffusion

Diffusion is the movement of ions or molecules from an area where they are highly concentrated to an area where they are in low concentration. This occurs in nonliving as well as living systems and is the result of the inherent nature of molecules to spread apart. This movement will continue until all of the molecules are equally dispersed and a state of equilibrium is reached. Unaided movement of this kind is known as **simple diffusion** (Figure 4-1). Molecules that are small and do not have a charge, such as O_2, CO_2, and H_2O, move across cell membranes in this manner (Figure 4-2). This type of movement is advantageous because it does not require the cell to use any energy.

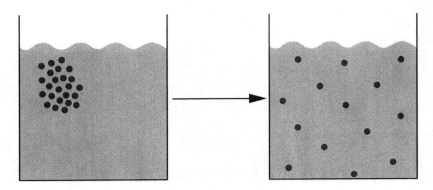

Figure 4-1 Simple Diffusion of a dye when put into water

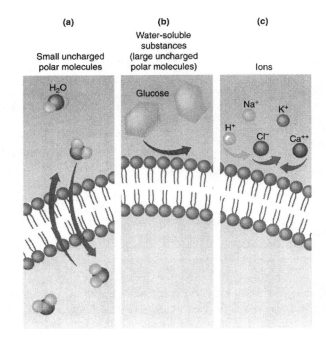

(a)

Small uncharged
polar molecules

H_2O

(b)

Water-soluble
substances
(large uncharged
polar molecules)

Glucose

(c)

Ions

Na^+ K^+

H^+

Cl^- Ca^{++}

Figure 4-2 Selective diffusion across a cell membrane.

Osmosis

The diffusion of water across a selectively permeable membrane is of such importance that it has been given its own name: **osmosis**. Osmosis will occur across cell membranes in response to the concentration gradient of water (high to low concentration) as well the concentration of dissolved particles in the water. Dissolved ions and molecules (**solutes**) are bound to water (**solvent**) molecules. Water molecules that are not bound to any solute are said to be "free." The concentration of dissolved particles, therefore, determines the concentration of "free" water in the solution. This phenomenon is known as **tonicity**: the relative concentrations of the water and the solutes on the outside and inside of cells (**Figure 4-3**). In living systems where molecules dissolved in water are abundant on the inside as well as the outside of cells, tonicity will have a great effect on osmosis. With respect to the concentration of dissolved particles inside of cells, three types of extracellular (outside of cell), solutions can exist (**Figure 4-4**).

Hypotonic Solution

A hypotonic solution has a lower concentration of dissolved particles than inside the cell. The solution will therefore have a higher concentration of free water molecules, and water will move into the cell.

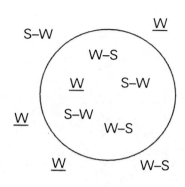
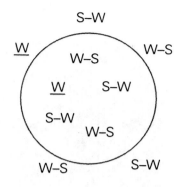
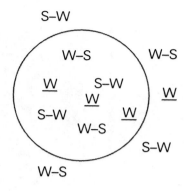

A. Hypotonic Solution   B. Isotonic Solution   C. Hypertonic Solution

Figure 4-3 This illustration depicts water (W) and solutes (S) in hypotonic, isotonic, and hypertonic solutions. The W's that are connected to the S's represent water that is bound to a solute. The W's that are underlined represent "free" water molecules that are able to move.

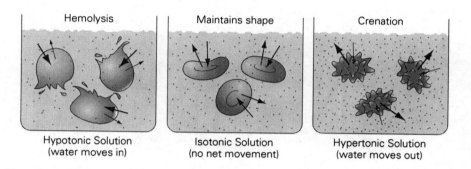

Figure 4-4 What happens to red blood cells when put into hypotonic, isotonic, and hypertonic solutions.

Isotonic Solution

An isotonic solution has the same concentration of dissolved particles and free water as the inside of a cell. Therefore, there will be no net movement of water into or out the cell.

Hypertonic Solution

A hypertonic solution has a higher concentration of dissolved particles than inside the cell. The cell will have a higher concentration of free water, and the water will exit the cell.

Facilitated Diffusion and Active Transport

Molecules that are too large, such as glucose ($C_6H_{12}O_6$), and atoms or molecules with charges (ions) cannot productively pass through the lipid bilayer cell membrane by way of simple diffusion. Proteins that act as "helpers" will "carry" these molecules from high to low concentration across cell membranes. Helpers are sometimes called "facilitators," and this phenomenon is known as **facilitated diffusion** (**Figure 4-5**). As with simple diffusion, this requires no energy expense (see Figure 4-2).

Sometimes physiological demands require molecules or ions to move against a concentration gradient: from low to high concentration. This is achieved through **active transport**. As with facilitated diffusion, a protein channel, usually called a pump, is employed . This action costs energy because ions or molecules are pushed in the opposite direction of which they would naturally move. Sodium ions are pumped in this way in the nervous system.

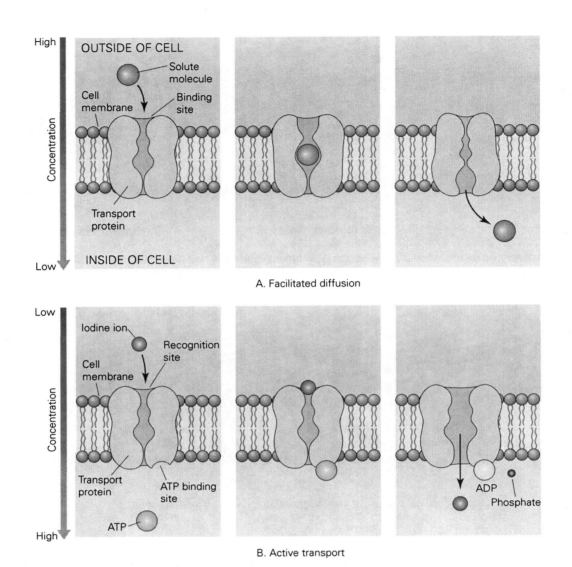

High

Concentration

Low

OUTSIDE OF CELL

Solute
molecule

Cell
membrane

Binding
site

Transport
protein

INSIDE OF CELL

A. Facilitated diffusion

Low

Concentration

High

Iodine ion

Recognition
site

Cell
membrane

Transport
protein

ATP binding
site

ATP

ADP

Phosphate

B. Active transport

Figure 4-5 **A.** Facilitated diffusion. **B.** Active Transport.

procedure

Diffusion

1. Fill a 500-ml beaker with tap water. Add several drops of methylene blue. Occasionally observe the dye in the water over a period of 30 minutes. While you observe the action of the dye, begin the osmosis exercise.

Osmosis: Eggs

2. Each group will be given a beaker with one of the sucrose solutions and a de-shelled chicken egg in the soaking solution.

   a. Remove the egg from the "soaking" solution, and carefully dry it with paper towels. Weigh the egg to the nearest 0.1 g. This will be recorded as the weight at time 0 minutes.

   b. Place the egg in the beaker of sucrose solution.

   c. At 15-minute intervals for 45 minutes, remove the egg from the sucrose solution. Wipe the egg dry, and weigh it.

   d. Record the results of every weighing in **TABLE 4-1**.

   e. Share data with other groups in the class and record the data in **TABLE 4-2**.

TABLE 4-1	Change in Weight of Egg Over Time	
	Weight of Egg	Change in Weight (+ or -)
0 min		————————
15 min		
30 min		
45 min		

TABLE 4-2	Weight Change for Every Sucrose Solution	
	Sucrose Solution	Change in Weight (+ or -)
0%		
10%		
20%		
30%		
40%		

Osmosis: Eodea Leaves—To Be Done While Waiting Between Egg Weighing Intervals

1. Prepare two wet mounts with elodea leaves: one using the isotonic solution in which the elodea is kept and the other using the 15% NaCl solution.

2. Observe both with the light microscope.

3 From your observations, determine the tonicity of the NaCl solution: hypotonic or hypertonic.

CLINICAL CONSIDERATIONS

Cystic Fibrosis

Cystic fibrosis is one of the most common inherited fatal diseases in the United States. It is caused by the failure to develop properly functioning protein pump channels in cell membranes that are responsible for transporting chloride ions (Cl^-). Properly functioning channels will transport Cl^- ions out of the cell into the extracellular fluid. Sodium (Na^+) and water are to follow the chloride ions based on a charge gradient and tonicity. When this chain of events fails to occur, the result is an overabundant production of mucus. This has a major impact on the respiratory system, as it obstructs the passageway for air.

Other systems that rely heavily on membrane transport, such as the digestive and urinary systems, will also be affected. The pancreas, which plays a major role in enzyme production and secretion in the digestive system, will eventually form "cysts." The cysts form in response to the mucus buildup, which blocks enzymes from leaving the pancreas.

Name: _____ Lab Section: _____

||||||| Review Questions

1. What is the term used to describe a solution that has the same concentration of water and particles that a cell has?

   _____

   _____

2. How does water cross a cell membrane?

   _____

   _____

3. How does oxygen cross a cell membrane?

   _____

   _____

4. In the lab procedure where you were weighing eggs, what was causing the weight changes?

   _____

   _____

5. Does osmosis require the use of energy?

   _____

   _____

6. Give an example of a molecule that crosses the cell membrane by way of facilitated diffusion.

   _____

   _____

7. How does sodium manage to move against a concentration gradient?

   _____

   _____

8. What would happen to one of your cells if it was put into pure water?

   _____

   _____

Name: _____ Lab Section: _____

9. Why is taking advantage of a concentration gradient beneficial for cells?

_____

_____

10. Which of the characteristics of living systems will be affected by membrane transport?

_____

_____

11. In the membrane transport exercise, which solutions were hypotonic? Which were hypertonic? Which solution, if any, was isotonic or close to it? How did you make these determinations?

_____

_____

Tissues: The Building Blocks of the Body

- To become familiar with the various tissue types in the body
- To be able to identify all of the tissue types under the microscope and state where they can be found
- To identify the layers and structures of the skin
- To gain an understanding of the levels of complexity of the body: cells, tissues, organs

- microscopes
- models of the skin

Slides of:

- nervous tissue
- connective tissue
  - areolar
  - adipose
  - dense
  - bone
  - cartilage
  - blood
- epithelial tissue
  - simple squamous
  - simple cuboidal
  - simple columnar
  - pseudostratified columnar
- muscle tissue
  - skeletal
  - cardiac
  - smooth

SAFETY ALERT!

Handle the microscopes carefully, and always carry them with two hands. Remember that the slides are made of glass. Even when not broken, the edges are sharp. In the event that a slide is broken, inform the instructor and seek guidance concerning cleanup procedures.

Introduction

A **tissue** is group of similar cells, including all of the extracellular (outside of the cell) material, organized to perform a specific function. The microscopic study of tissues is know an as **histology**. As microscopes have become more sophisticated, the science of histology has been able to reveal similarities and differences between tissue types. In humans, and vertebrates in general, four basic tissue types exist: **epithelial, muscle, nervous,** and **connective**.

procedure

Studying histology can be quite challenging and frustrating. Many of the slides you examine will look very similar upon first inspection. This is a "visual" exercise. With time and experience using the microscope, you will become proficient. Consult your instructor and lab manual as often as needed.

Observe the figures, diagrams, and plates of the tissues.

1 After you have a clear idea of what you might see, observe slides of theses tissues under the microscope.

2 As when you studied cells, begin on the lowest power (scanning).

3 Using skills that you have developed, view the tissues on higher powers.

4 Space is provided at the end of the exercise to make drawings of the tissues.

5 Review the locations and functions of these tissues.

Epithelial tissue

Epithelial tissue is found lining organs, vessels, and body cavities. Here it forms **membranes**, or sheets of cells lying on top of connective tissue, where it functions to provide **protection** and **support,** such as in the skin.

Some epithelial tissue, such as that found in the digestive tract, is involved in **absorbing** nutrients to pass into the bloodstream.

Glands, which produce products to be **secreted** into various parts of the body, are primarily built from epithelial tissues. These include the pancreas and the salivary glands.

Epithelial tissues consist of cells tightly bound together with very little extracellular material. They are also "**avascular.**" That is, blood vessels do not penetrate into the tissue.

Epithelial tissue is categorized and named based on the shape of the cells and how many layers of cells are present. Epithelial cells can be classified as **squamous, cuboidal,** or **columnar.** Squamous refers to cells that are somewhat oval or shapeless. No geometric shape can be ascribed to them. Cuboidal cells, as the name implies, are roughly shaped like cubes. That is, they are about as tall as they are wide. Columnar cells (columns) are taller than they are wide.

You will not always be able to identify the cell membrane borders. Therefore, studying the arrangement of the nuclei will allow you to determine the shape.

Epithelium can also be classified as either **simple (one-layer thick)** or **stratified (comprised of two or more layers).** Names for epithelial tissues will include terms that describe the cell shape and layering. For example, if the tissue contains one layer of cube-shaped cells, it will be called "simple cuboidal epithelium."

Epithelial tissue will have a **free surface** where the outermost layer of cells is in contact with other cells or tissues. The bottom layer is attached to the basement membrane (**Figure 5-1**). The basement membrane is a specialized protein that is secreted by epithelial cells and usually attached to underlying connective tissue.

Simple Epithelial Tissues

Squamous (**Figure 5-2**). This type of epithelium is found lining vessels and ducts. Its functions include diffusion and osmosis.

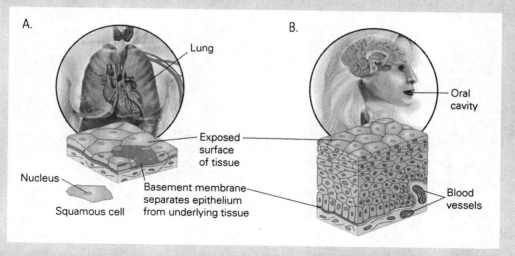

Figure 5-1 A. Simple squamous epithelium. B. Stratified squamous epithelium.

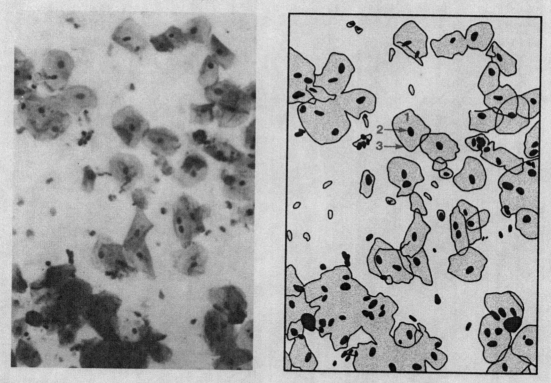

Figure 5-2 Simple squamous epithelium. 1. cytoplasm. 2. nucleus. 3. cell membrane.

Cuboidal (Figure 5-3). This can be found in the lining of the kidney tubules where it engages in filtration.

Columnar (Figure 5-4). This is found in the lining of the digestive tract, especially the intestines where it has secretory and absorptive functions. *Locate a goblet cell.

Stratified squamous epithelium (Figure 5-5). The classic example of stratified squamous epithelium is the **epidermis of the skin**. It serves as a protective barrier to the outside world. *Note the many layers between the free surface and the basement membrane.

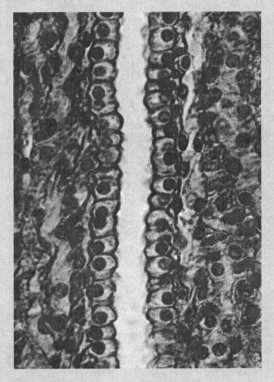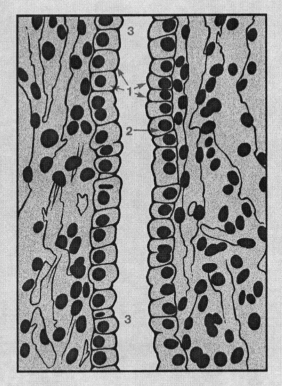

Figure 5-3   Simple cuboidal epithelium. 1. cuboidal cells. 2. nucleus of cuboidal cell. 3. lumen of tubule.

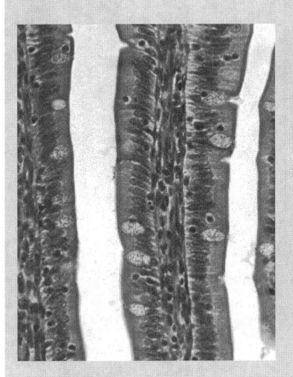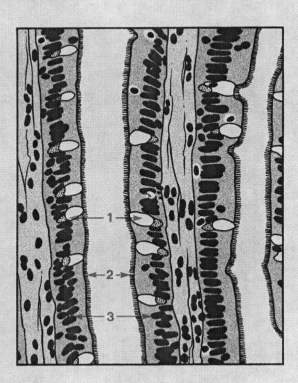

Figure 5-4   Simple columnar epithelium. 1.goblet cells. 2. brush border. 3. nuclei of columnar cells.

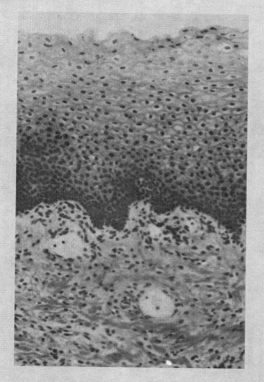 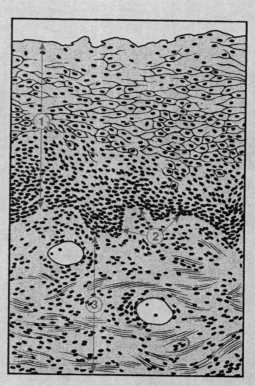

Figure 5-5 Stratified squamous epithelium; layers of flattened cells serve to protect underlying tissues. 1. stratified squamous epithelium. 2. basement membrane. 3. loose connective tissue.

Pseudostratified columnar epithelium (Figure 5-6). Pseudostratified means "falsely layered." This type of tissue can appear stratified; however, after close inspection, all of the cells actually touch the basement membrane. It can be found in the respiratory tract where it aids in the movement of mucus.

Muscle tissue

Muscle tissue is unique in that it can generate a force. Three types of muscle tissue exist: **skeletal, cardiac,** and **smooth.**

Skeletal muscle (**Figure 5-7**) is the type that is most often referred to when using the term muscle. It is fast acting, under **voluntary control,** and brings about movement of the skeleton as in walking or chewing. Under the microscope, skeletal muscle appears striped, or **striated.** This appearance is due to the highly organized alternating bands of the protein fibers actin and myosin. Skeletal muscle cells, or **fibers,** each have many nuclei.

Cardiac muscle (**Figure 5-8**) is restricted to the heart. It also has a striated appearance and is fast acting. It is, however, under **involuntary control.** That is, the nervous system dictates its rhythm without conscious input. The fibers are somewhat branched and usually have only one nucleus. **Intercalated discs** hold the fibers tightly together. This is to ensure that the signals from the nervous system are quickly communicated to all cardiac muscle cells. ***Cardiac muscle and skeletal muscle appear very similar.** Intercalated discs and the branching of cardiac muscle fibers can be used to distinguish the two.

Smooth muscle (**Figure 5-9**) is under involuntary control and is found in many places such as the digestive tract, blood vessels, and ducts that control the passage of glandular products. The name smooth is derived from the fact that under the microscope it has a smooth, nonstriated

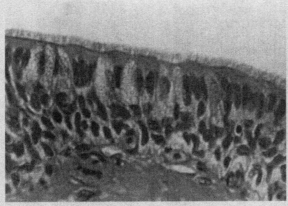 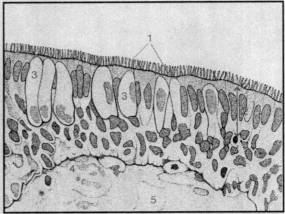

Figure 5-6 Pseudostratified ciliated columnar epithelium; this tissue type is found primarily in the respiratory tract. 1. cilia. 2. pseudostratified columnar cells. 3. goblet cell. 4. basement membrane. 5. loose areolar connective tissue.

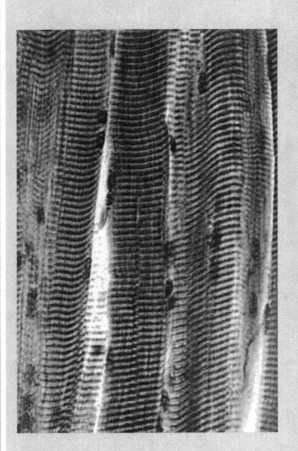 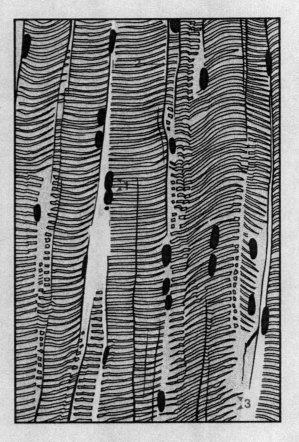

Figure 5-7 Skeletal muscle; fibers (cells) appear striated with peripheral oval nuclei. Comprises voluntary muscles which move bones. 1. nucleus. 2. skeletal muscle fiber. 3. striations.

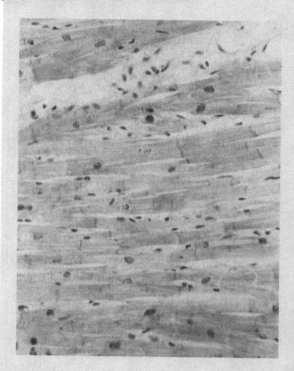
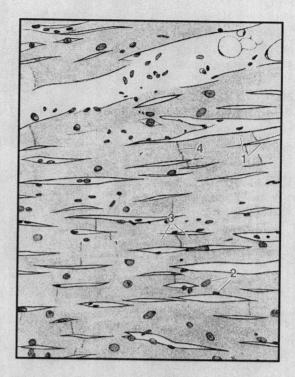

Figure 5-8 Cardiac muscle is a type of striated muscle that contains bifurcations and centrally located nuclei. 1. bifurcations. 2. nuclei. 3. striations. 4. intercalated disks.

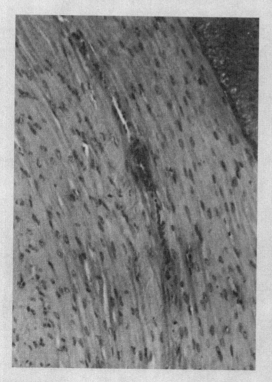
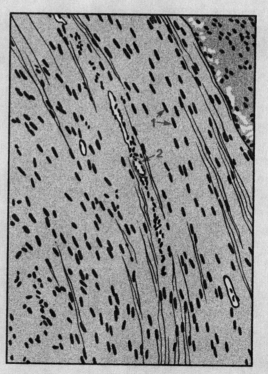

Figure 5-9 Smooth muscle; this tissue type is found in blood vessels and in the digestive tract and is under involuntary nervous control. 1. nuclei of smooth muscle fibers. 2. capillary.

appearance because the protein fibers are not as organized. The contractions generated from smooth muscle tend to be slow and prolonged. That is, as opposed to cardiac and skeletal, it is slow acting. Smooth muscle cells have only one nucleus.

Nervous Tissue

Nervous tissue is found in the brain, spinal cord, and all of the nerves that supply all organs and body regions. It is comprised of two types of cells: **neurons** and **neuroglial** cells. Neurons are unique in that they conduct signals, or electrical impulses, very rapidly. They also have a very distinct structure. Three regions can be identified on them: **dendrites, the cell body**, and the **axon** (**Figure 5-10**). Neuroglial cells function as specialized connective tissue. They lend support and protection to all structures in the nervous system.

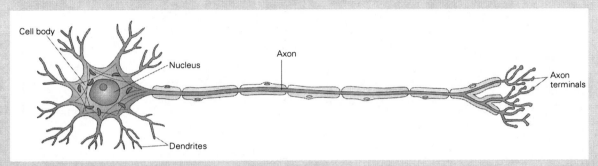

Cell body

Nucleus

Axon

Axon terminals

Dendrites

Figure 5-10 A typical neuron.

Connective Tissue

There are many diverse types of connective tissues; however, they all share common traits. They have far fewer cells than the other tissues previously discussed. The cells are rarely in contact with each other and are surrounded by an extracellular material known as the **matrix**. The matrix contains **protein fibers** in varying amounts and can range from a solid, such as the calcium phosphate crystals in bone, to the fluid found, the plasma of blood.

Connective tissues have historically been classified in several ways. In this exercise, we recognize four subcategories: **connective tissue proper, bone, cartilage,** and **blood**.

Connective Tissue Proper

Areolar tissue is a "loose" connective tissue. There is plenty of space between the cells and the fibers are not tightly bound (**Figure 5-11**). It is soft and pliable and can be found directly under the skin and covering muscles and other organs. *Identify and distinguish between the thick collagen and thin elastic fibers. Are the fibers tightly bound together?

Adipose is also a loose connective tissue (**Figure 5-12**). It is an exceptional connective tissue in that it has a high density of cells with very little associated matrix. The cells are called **adipocytes** and are filled with fat droplets. Adipose is found throughout the entire body where it provides insulation and cushioning and serves as a storehouse for energy. *The cells are close together, and the nuclei are pressed tightly against the cell membrane because the cells are filled with fat.

Dense connective tissues are characterized by a high density of protein fibers (**Figure 5-13**). These tissues are built to resist forces while remaining somewhat elastic. They are found in **tendons** and **ligaments**. *The fibers are tightly bound together.

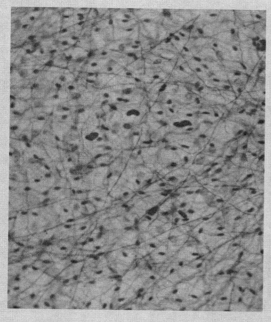

Figure 5-11 Loose areolar connective tissue; contains cells and fibers. Functions as binding and packing substance around organs. 1. elastic fibers. 2. collagenous fiber. 3. macrophage cell. 4. fibroblast cells.

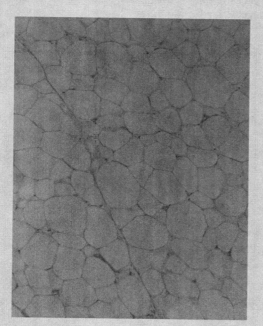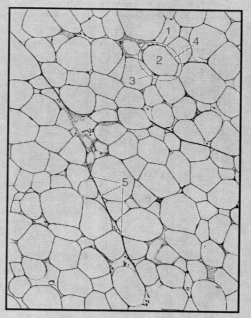

Figure 5-12 Adipose tissue is composed of adipocytes containing fat. 1. adipocyte (adipose cell). 2. fat vacuole. 3. nucleus. 4. cell membrane. 5. capillary.

Bone

Bone is built from a very hard calcium phosphate matrix. The matrix is secreted by **osteocytes**, or bone cells (**Figure 5-14**). Osteocytes are organized in rings around a **central (haversian)** canal that contains blood vessels and nerves. Little canals, **canaliculi**, carry nutrients and waste between the osteocytes and central canal. ***The osteocytes are arranged around the central canal. Locate canaliculi and the matrix.**

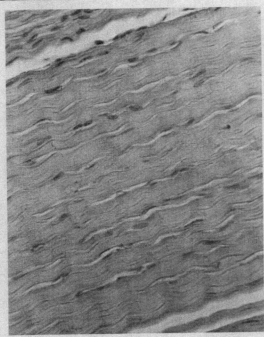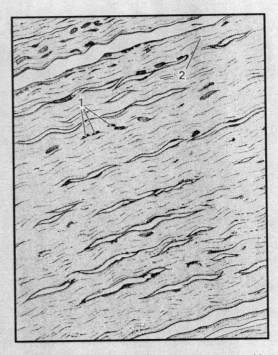

Figure 5-13 Dense, fibrous connective tissue (regular); closely aligned parallel rows of wavy fibers are separated by single rows of darker staining cells; found in tendons and ligaments. 1. cells (fibrocytes). 2. fibers.

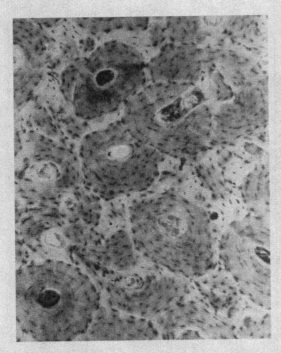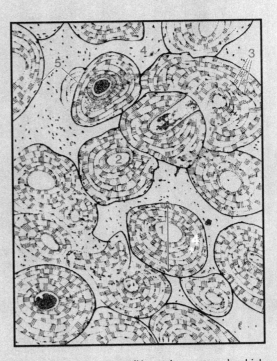

Figure 5-14 Ground bone (compact bone); this type of osseous tissue contains osseons (Haversian systems), which contribute to its rigidity. 1. osteon (Haversian system). 2. osteonic canal (haversian canal). 3. osteocytes. 4. matrix. 5. canaliculi.

Cartilage

Cartilage is not as rigid as bone, but is more flexible. The cells in cartilage, known as **chondrocytes**, are also surrounded by a solid matrix (**Figure 5-15**). The most common type is hyaline cartilage. It is found in the nose, joints, and connection between the sternum and ribs. *The chondrocytes are not as organized as the osteocytes in bone.

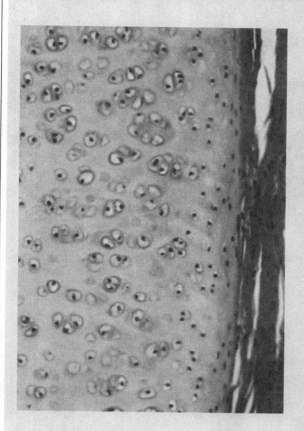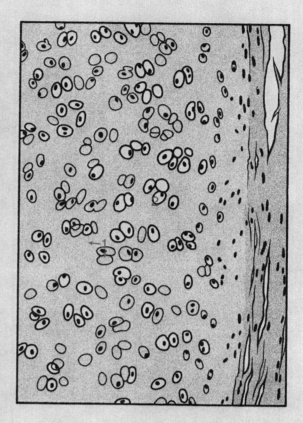

Figure 5-15 Hyaline cartilage of the trachea; when stained, this tissue type has a bluish, glasslike appearance; also found covering ends of bones to reduce friction. 1. chondrocytes. 2. intercellular matrix.

Blood

It may be hard to think of blood as a connective tissue; however, it meets the criteria. It has cells and a matrix. **Red blood cells** and **white blood cells** are carried in a fluid matrix called plasma (**Figure 5-16**). Blood is found in the heart and all of the blood vessels and functions as transport and defense systems. ***Be able to distinguish red blood cells and white blood cells.**

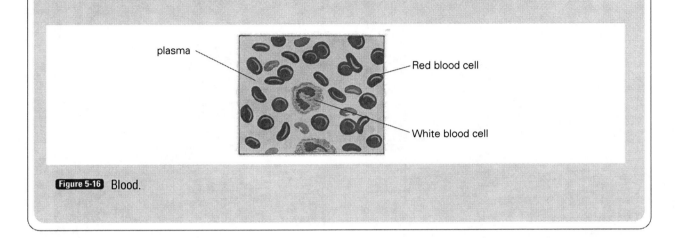

Figure 5-16 Blood.

‖‖‖‖ From Tissues to Organs

An organ is comprised of two or more tissue types brought together to perform specific functions. The **skin** is an organ that is made up of all four basic tissue types (Figure 5-17). The top layer, the **epidermis**, is a stratified squamous epithelium as mentioned above. The outer layers of cells are filled with the protein **keratin** and are actually dead. They serve to prevent water loss and to protect from harmful environmental factors such as bacterial invasion. These outer layers are continually shed. The bottom layer of the epidermis is mitotic and gives rise to new cells.

The lower layer of the skin is the **dermis**. It is a sea of loose connective tissue containing many structures. Found in the dermis are nerves, muscles, vessels, glands, and hair. The glands and hair are actually epidermal derivatives. That is they are built from epithelium but invaginate down into the dermis.

The skin functions in protection, temperature regulation, sensation, and synthesis of vitamin D.

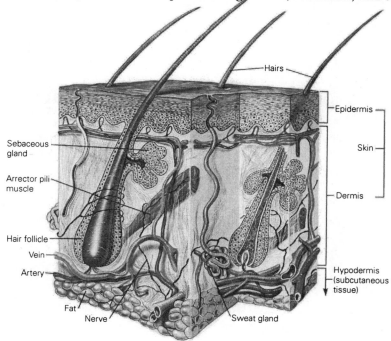

Figure 5-17

CLINICAL *CONSIDERATIONS*

Stem Cells

All of the mature tissues that we observed in this exercise originally had arisen from stem cells. Stem cells are early embryonic, undifferentiated cells. They possess the ability to become many types of specialized cells and tissues.

Stem cells can be found throughout life in the epithelial lining of the skin and digestive tract where cells are constantly being shed and replaced.

Stem cells also exist in bone marrow where they continually form new red and white blood cells.

Stem cells can be used to replace and repair tissues for the treatment of cancer, Parkinson's disease, Alzheimer's disease, diabetes, and arthritis. Reservoirs of adult stem cells are too small and difficult to access. Therefore, research has focused on ethical methods to acquire large quantities of embryonic stem cells.

Name: _____ Lab Section: _____

|||||| Review Questions

1. Which type of tissue is characterized by tightly bound cells?

   _____

   _____

2. Which connective tissue has a calcium phosphate matrix?

   _____

   _____

3. Where in the body can dense connective tissue be found?

   _____

   _____

4. What is the formal name for cartilage cells?

   _____

   _____

5. Which type of muscle is fast acting and under involuntary control?

   _____

   _____

6. In order to make you bleed a cut has be deep enough to make it into what layer of skin?

   _____

   _____

7. Which type of muscle is under involuntary control and is not striated?

   _____

   _____

8. Where is simple cuboidal epithelium found?

   _____

   _____

9. What is the matrix called in blood?

   _____

   _____

Name: _____ Lab Section: _____

10. What allows for communication between the osteocytes and the central canal?

_____

_____

11. List all of the functions of adipose.

_____

_____

12. Draw and label the structure of a neuron.

13. Draw and label the structure of the skin.

Name: _____   Lab Section: _____

Drawings of Tissues:

Name: _____ Lab Section: _____

Digestive System 1: Anatomy

◤ SAFETY ALERT!

Preserved specimens contain chemicals that are potentially irritating to the skin and eyes. Do not handle specimens without eyewear or gloves. Dissection instruments are sharp. Take care not to puncture or cut your skin.

||||||| Introduction

Digestion is the process by which food is transformed into **usable nutrients** to be absorbed into the bloodstream. Ingested carbohydrates, proteins, and fats are broken down into the usable nutrients glucose, amino acids, and fatty acids, respectively.

Two components to digestion exist: **mechanical** and **chemical**. Mechanical digestion includes chewing, swallowing, and pushing the food through the digestive tract. Chemical digestion mostly involves **enzymatic action** to finalize the breakdown process.

In this exercise, you will study the anatomical structures involved in the digestive process. The **digestive tract proper**, or **alimentary canal**, is the tube where the food actually passes through. It consists of the **mouth, pharynx, esophagus, stomach, small intestines, large intestines, rectum,** and **anus** (Figure 6-1).

Chemicals that aid in the digestive process are produced by accessory organs and secreted into the digestive tract. These accessory organs, which include the **salivary glands, liver, gall bladder,** and **pancreas,** are never in contact with the food.

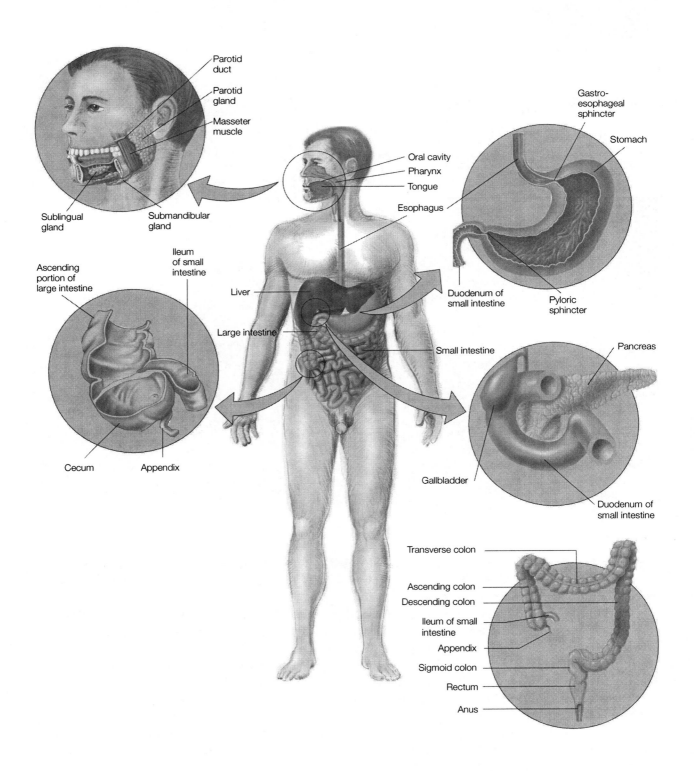

Figure 6-1 The Human Digestive System.

Parotid duct

Parotid gland

Masseter muscle

Sublingual gland

Submandibular gland

Oral cavity

Pharynx

Tongue

Esophagus

Gastro-esophageal sphincter

Stomach

Duodenum of small intestine

Pyloric sphincter

Ileum of small intestine

Ascending portion of large intestine

Liver

Large intestine

Small intestine

Pancreas

Cecum

Appendix

Gallbladder

Duodenum of small intestine

Transverse colon

Ascending colon

Descending colon

Ileum of small intestine

Appendix

Sigmoid colon

Rectum

Anus

procedure

Using the diagrams in the manual and the models of the human torso, identify all of the structures of the digestive system listed here.

1 Mouth, teeth, and salivary glands

Humans have an "omnivorous" diet. That is, they eat many types of foods. This is reflected in the structures of the teeth used for chewing (**Figure 6-2**). **Incisors** and **canines** are for ripping and tearing meats, whereas **premolars** and **molars** are built to grind fruits and vegetables.

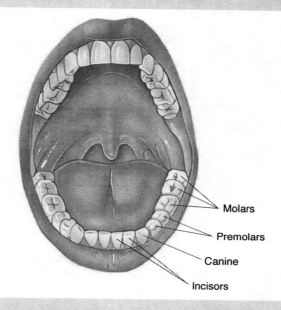

Molars

Premolars

Canine

Incisors

Figure 6-2 The Teeth. Adults have 32 teeth.

Although only in the mouth for a short period, here the chemical assault on food begins. The **salivary glands (parotid, sublingual,** and **submandibular)** produce saliva. It is comprised mostly of water but also contains the enzyme **amylase.** Amylase begins the chemical digestion of starches by breaking them into the disaccharide **maltose.**

The tongue then pushes the food into the portion of the oral cavity known as the **pharynx.** The act of swallowing sends the food into the "rubbery" tube called the **esophagus.**

2 Esophagus and stomach

The esophagus is located behind the trachea and extends beyond the heart and lungs to terminate at the stomach. It serves only to transport food. This is accomplished by way of smooth muscle contractions known as **peristalsis.** While in the esophagus, no chemical digestion of the food occurs.

The **stomach** is a J-shaped organ lying below the heart and above the intestines on the left side of the body. The wall of the stomach is comprised of three layers of smooth muscle. Contractions of the muscular wall allow for mechanical digestion by churning the food. This mixes the food with the highly acidic **gastric juice.** Gastric juice contains the enzyme **pepsin,** which begins the chemical digestion of proteins by breaking peptide bonds.

The food is now an acidic ball called **chyme.** Through further peristaltic action, it is pushed into the first part of the small intestine, the **duodenum.**

3 Small intestines, liver, gall bladder, and pancreas

The duodenum arises from the right side of the stomach and curves downward. It makes up only the first 10 inches of the 17-feet long small intestine. However, it is a region of intense chemical activity. The pancreas, liver, and gall bladder all surround the duodenum (Figure 6-3). These accessory organs release fluids into the duodenum that aid in chemical digestion.

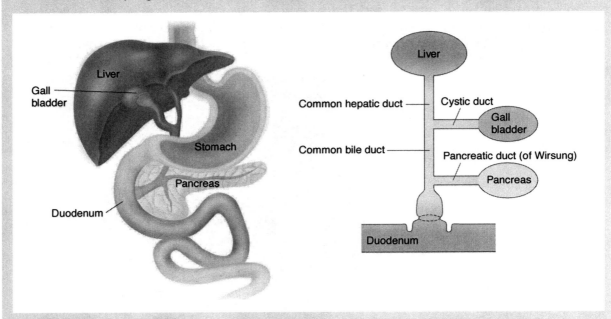

Figure 6-3 The liver, gall bladder, pancreas, and duodenum are interconnected as shown. The schematic diagram makes this anatomy easier to comprehend.

4 Liver and Gall Bladder

The **liver** is a large glandular organ that lies just below the **diaphragm**, mostly under the ribs (thoracic cavity). It extends from the stomach onto the right side of the cavity. The liver has several functions. It produces **bile** that is stored in the **gallbladder**. The gall bladder is a small "greenish" little sac that lies just below, and is actually an outgrowth of, the liver.

Bile is released from the gall bladder into the duodenum through the **common bile duct**, where it **"emulsifies"** fats. That is, it breaks large fat molecules into smaller droplets. This increases the surface area for the action of enzymes.

The liver also serves as a filter through which most newly absorbed nutrients pass. The **hepatic portal vein** carries blood from the digestive tract to the liver before it reaches the general circulation (Figure 6-4). This provides a preliminary screening for any toxic materials that might have been ingested.

Finally, the liver is involved in breaking down and recycling worn out red blood cells and producing plasma proteins.

5 Pancreas

The **pancreas** is also a glandular organ. It arises from the inner curvature of the duodenum and has two distinct portions. The **head** follows the contour of the duodenal curvature, and the **tail** extends toward the midline of the abdominal cavity.

As an **exocrine** (outside of the blood stream) gland, the pancreas secretes pancreatic juice

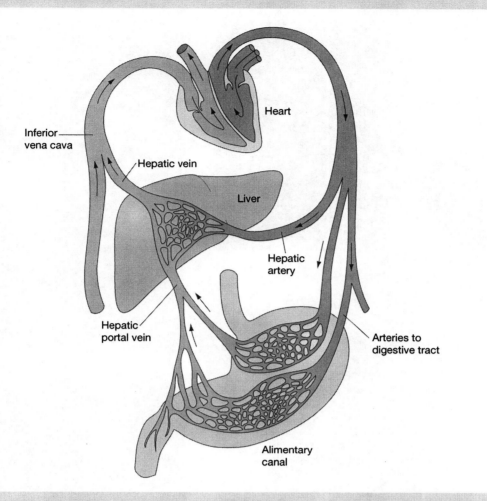

Figure 6-4 The hepatic portal system.

into the duodenum. It contains enzymes that break down starches, proteins, and fats and a buffer that neutralizes the acidic chyme.

As an **endocrine** (inside the blood stream) gland, the pancreas produces **insulin** and **glucagon**. These two hormones work together to maintain proper blood glucose levels.

6 Jejunum and Ileum

The remainder of the small intestine includes the 6-feet section called the **jejunum** and the 10-feet section called the **ileum**. There is no anatomical landmark that divides the two.

Nutrients will move slowly through these regions by way of peristalsis. These last regions of the small intestine are lined with simple columnar epithelium that is specialized for absorbing nutrients. The surface for absorption is increased by villi (**Figure 6-5**). The remainder of the "food" is now moved into the large intestine.

7 Large intestine, rectum, and anus

At only 4 to 5 feet long, the large intestine is much shorter than the small intestine (**Figure 6-6**). It is termed "large" because it is larger in diameter. The small intestine meets the large intestine at a region called the **cecum**. The **appendix** is found as a small thin extension of the cecum.

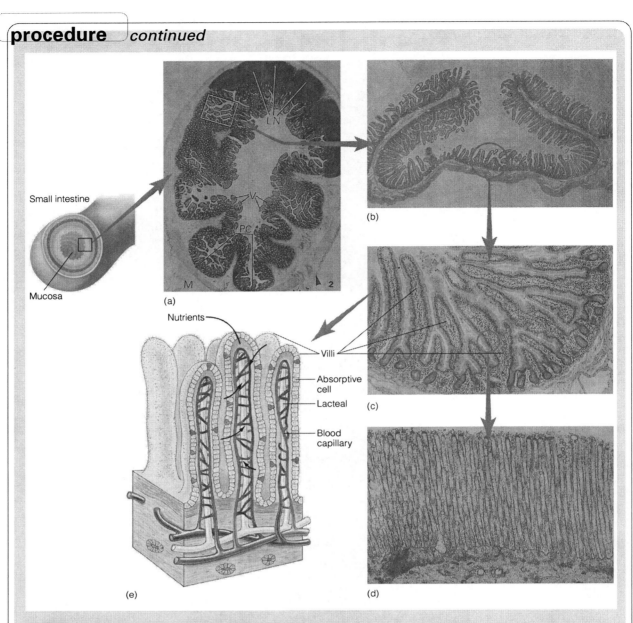

Figure 6-5 The Small Intestine. The small intestine is uniquely "designed" to increase absorption. (a) A cross section showing the folds. (b) A light micrograph of folds and villi. (c) Higher magniifcation of villi. (d) A electron micrograph of the surface of the absorptive cells showing the microvilli. (e) Each villus contains a loose core of connective tissue; a lacteal, or lymph, capillary; and a network of blood capillaries. Nutrients pass from the lumen of the small intestine through the epithelium and into the interior of the villi, where they are picked up by the lymph and blood capillaries.

The first part of the large intestine extends upward in the abdominal cavity and is called the **ascending colon**. From there the intestine bends at a 90-degree angle to become the **transverse** colon that extends to the left side of the abdominal cavity. It makes another 90-degree turn and extends downward as the **descending colon**. The final portion curves back toward the midline and is known as the **sigmoid colon**. From there arises the rectum. The tract ends with the anus.

The large intestine is involved in feces formation and the reabsorption of water. It also finalizes

procedure *continued*

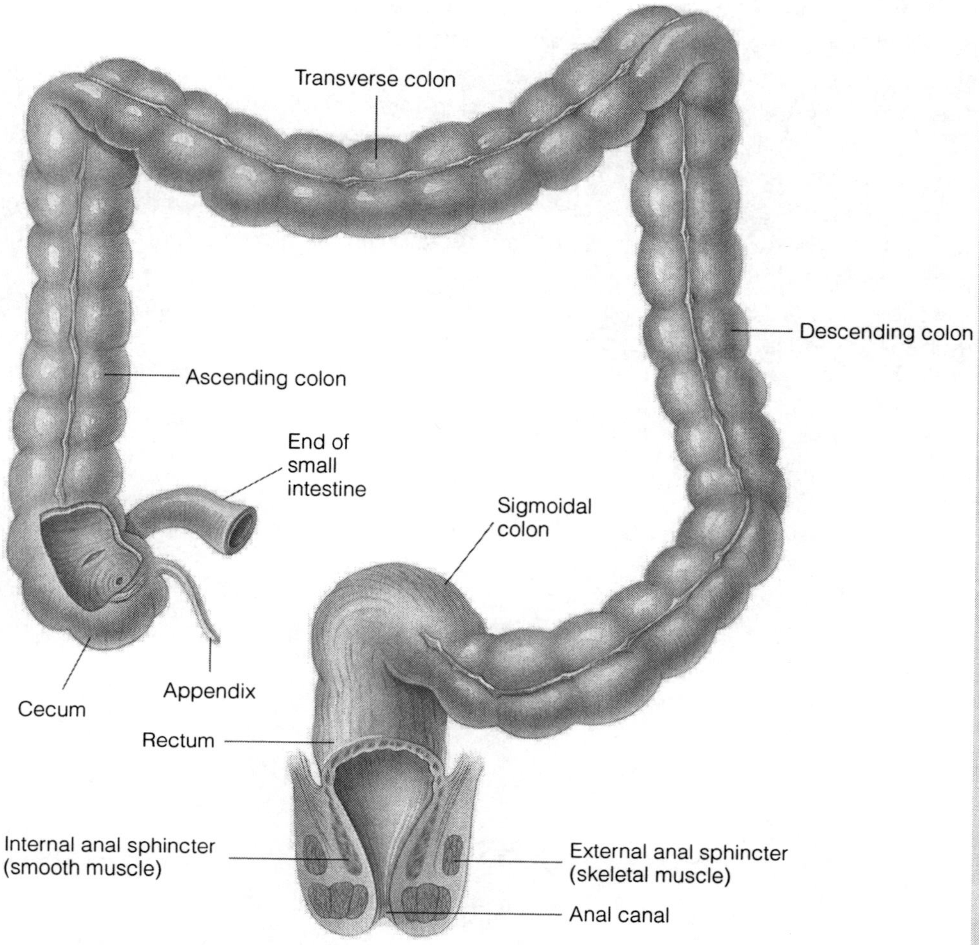

Figure 6-6 The large intestine.

carbohydrate digestion through the aid of the naturally occurring bacteria ***E. coli.***

Elimination of waste, or **defecation**, involves muscular contractions of the rectum and anus.

8 Slide of digestive tract lining

a. Observe all available slides of the digestive tract lining under low and high powers on the microscope.

b. Note the differences in the epithelial, connective tissue, and smooth muscle layers from the various regions.

9 Dissection of preserved specimens

As this will be your first exercise using dissection, consult the instructions in the beginning of Appendix 2. Follow the instructions guiding you on proper techniques for making incisions to reveal the internal organs. Locate all of the structures labeled in the Appendix.

Colorectal Cancer

Colorectal cancer is the second most common cancer in males and the third in females. Roughly half of all instances of this cancer can be linked to a genetic component; however, diet and life style play a contributing role. A diet that is high in animal fat and consumption of alcohol has also been linked to an increased risk of developing this cancer.

Symptoms may include constipation alternating with bloody diarrhea. Unfortunately, these symptoms may not occur until the cancer is quite advanced, lowering chances of recovery. It is recommended that people who are over the age of 40 years and who have a family history of this cancer be checked yearly. This includes a rectal exam, examination of feces for blood content, and a colonoscopy.

Diets that are high in fibers that can be found in fruits, vegetables, and some grains are thought to reduce the risks of colorectal cancer.

Name: _____ Lab Section: _____

IIIIII Review Questions

1. Where in the digestive tract are proteins first chemically digested?

   _____

   _____

2. Which organ stores bile?

   _____

3. Which nutrient is emulsified by bile?

   _____

4. Which region of the small intestine is the longest?

   _____

   _____

5. Which part of the alimentary canal has three layers of smooth muscle?

   _____

   _____

6. Which nutrient is chemically digested in the esophagus?

   _____

7. Which vessel carries blood rich in nutrients from the digestive tract to the liver?

   _____

8. Discuss the endocrine and exocrine roles of the pancreas.

   _____

   _____

9. What is meant by "omnivorous" diet? Which anatomic structures could you study to
   determine diet?

   _____

   _____

10. Name the three salivary glands and describe their anatomic location.

   _____

   _____

Digestive System 2: Physiology

objectives

- To investigate the effects of enzymes on starches, fats, and proteins
- To investigate the effects of temperature and pH on enzyme action

materials

- albumin solution
- pancreatic lipase
- starch solution
- phenol red
- maltose solution
- bile salts
- iodine
- Benedict's solution
- Biuret reagent
- 500-ml beakers
- test tubes and racks
- hot plate
- potatoes
- 0.03% HCl
- knives or razor blades

▼ SAFETY ALERT!

Many of the chemicals that are used in this exercise are either caustic or toxic or both. Take care not to spill any of these on your clothes or skin. The hot plates and water baths are to be regarded as potentially dangerous because of the heat. You will be using glassware (test tubes and beakers). In the event that any glass is broken or you get cut with a knife or razor blade, consult your instructor concerning first aid and cleanup procedures.

||||||| Introduction

Mechanical digestion, such as chewing, increases the surface area on which chemical digestion can occur. Most chemical digestion is through enzymatic action.

Enzymes are highly specialized **catalysts**. Without the aid of enzymes, chemical reactions could never proceed efficiently enough to sustain living systems. Digestive enzymes break down large macromolecules into smaller "nutrients" that can easily be absorbed into the bloodstream (**Figure 7-1**).

Enzyme action increases with temperature. However, extreme heat will denature enzymes. That is, their unique three-dimensional shape will be destroyed, rendering them ineffective. Enzymatic action is also sensitive to changes in **pH**.

In this exercise, you will study the effects of the enzymes **amylase**, **lipase**, and **pepsin** on starch, fats, and proteins, respectively. You will also investigate enzymatic action with varying temperatures and pH.

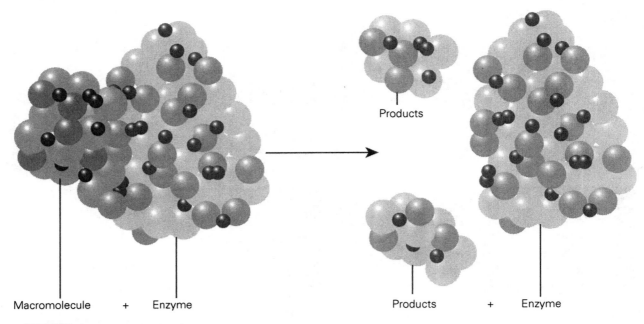

Macromolecule + Enzyme

Products

Products + Enzyme

Figure 7-1 Enzyme action in digestion.

procedure

1 Digestion of starch using amylase

Starch is the complex carbohydrate found in grains, potatoes, pasta, and breads. It is a polysaccharide that is made of many glucose molecules. Glucose is the usable nutrient that can be absorbed into the bloodstream. The first step in chemically degrading starch is to break it down into the disaccharide maltose. This is accomplished through the action of the enzyme **salivary amylase** in the mouth:

$$\text{starch + water} \xrightarrow{\text{Amylase}} \text{maltose}$$

Two tests for starch digestion will be run (i.e., tests for the action of amylase). If amylase has not acted, the starch solutions will turn blue, black, or gray in the presence of iodine. If it has acted, the solution should contain maltose and will turn from green to red in the presence of Benedict's solution. Green indicates low concentration and red indicates high concentrations.

  a. Fill two 500-ml beakers with 250-ml water each, and heat both on hot plates. These will serve as the water baths.

  b. Label eight test tubes with numbers 1 through 8. In steps 3 and 4 you will prepare controls.

  c. Fill test tubes 1 and 2 with a few drops of iodine. Add 5 ml of starch solution to tube 1 and 5 ml of distilled water to tube 2. Starch should turn blue, gray, or black in the presence of iodine. Record color changes:

  Tube 1_____

  Tube 2_____

d. Fill test tubes 3 and 4 with a few drops of Benedict's solution. Add 5-ml maltose solution to tube 3 and 5-ml distilled water to tube 4. Put both tubes into a boiling water bath for 10 minutes. The maltose solution should turn green or red in the presence of the Benedict's solution. Record the color changes:

Tube 3_____

Tube 4_____

e. Fill test tubes 5 and 6 each with 5-ml amylase solution and 5-ml starch, and let them stand for 20 minutes. Fill 2 test tubes 7 and 8 with 5-ml amylase solution, and place them into the boiling water bath for 10 minutes. Remove them from the water bath, and let them stand for 20 minutes.

f. Add 5 ml of Benedict's solution to tubes 5 and 7, and put them into the boiling water bath for 10 minutes to test for maltose. Add 5-ml iodine solution to test tubes 6 and 8 to test for the presence of starch. Record your results.

Tube 5_____

Tube 6_____

Tube 7_____

Tube 8_____

What differences did you expect to find in tubes 5 and 6 and tubes 7 and 8? What did you actually find? Explain your expectations and findings.

What did this exercise illustrate about enzyme action?

Answer:_____

2 Digestion of fats using bile and lipase

Digestion of fats is a two-step process in the small intestine. Bile will first mechanically digest, or emulsify, the large fat molecules into smaller droplets. This increases the surface area for the action of the enzyme **pancreatic lipase**:

$$\text{fat droplets} + \text{water} \xrightarrow{\text{Lipase}} \text{fatty acids and glycerol}$$

You can indirectly measure the action of lipase by using the pH indicator **Phenol Red**. It is red in a basic environment but will turn yellow in acidic conditions. As lipase acts, the conditions become acidic.

a. Fill three test tubes each with 5-ml vegetable oil and 10-ml Phenol Red.

b. In test tube 1 add 5-ml pancreatic lipase (pancreatin solution) and a pinch of bile salts.

In test tube 2 add 5-ml pancreatic lipase. In test tube 3 add 5-ml distilled water. Swirl the tubes gently to mix the solutions and record the colors.

Tube 1_____

Tube 2_____

Tube 3_____

c. Put all three tubes into a water bath that is kept between 30°C to 40°C. Check the tubes every 15 minutes for any color change. Note how long it takes to see a change in color in all three tubes.

Tube 1_____

Tube 2_____

Tube 3_____

What conclusion can you draw about fat digestion and surface area?

Answer:_____

3 Digestion of proteins using pepsin

Protein is found in foods such as meat, beans, and eggs. In this exercise you will use **albumin**, the protein found in eggs, to investigate the action of the enzyme **pepsin**. In the stomach, pepsin begins the chemical digestion of proteins by breaking peptide bonds:

$$protein + water \xrightarrow{Pepsin} peptides$$

The action of pepsin is dependent on the highly acidic (low pH) environment found in the stomach.

Biuret reagent will turn purple in the presence of whole proteins; however, it will turn a pinkish purple if peptides are present. Therefore, it can be used as indicator for the action of pepsin.

a. Label and fill three test tubes each with 5-ml albumin solution.

b. In tube 1, add 5-ml pepsin solution and 5-ml 0.03 HCl. Mix gently and place into a water bath set at 35°C to 40°C for 30 minutes.

c. Into tube 2, add 5-ml pepsin solution and put into the water bath for 30 minutes.

d. Into tube 3, add 5-ml distilled water and put into the water bath for 30 minutes.

e. Record any color changes:

Tube 1_____

Tube 2_____

Tube 3_____

What can you conclude about enzyme action and pH?

Answer:_____

4 Increasing surface area for digestion

a. Cut two small pieces of potato, each about 1 cm wide and 2 cm long.

b. Place the one of the pieces into test tube 1, and add 5-ml amylase solution.

c. Crush the other piece using a mortar and pestle, and scoop it into test tube 2.

d. Heat both tubes in a "warm" water bath (about 30°C to 40°C) for 15 minutes.

e. Add 5-ml Benedict's solution to each and put into water bath for 10 minutes.

f. Record the changes in color:

Tube 1_____

Tube 2_____

What can you conclude about surface area and chemical digestion?

Answer:_____

Common Physiological Disturbances in Digestion

Lactose intolerance is caused by insufficient, or lack of, production of the enzyme lactase. It is produced in the lining of the small intestine and is responsible for the chemical breakdown of lactose, the disaccharide sugar found in dairy products. Lactase degrades lactose into the monosaccharides glucose and galactose, both of which can easily be absorbed into the bloodstream. Undigested lactose causes fluid to be retained in the feces and fermentation leading to an excess production of gases. The result is usually cramping and diarrhea. Treatment involves avoidance of high dairy intake supplemented with tablets that contain lactase taken before meals.

Overproduction of HCl (hydrochloric acid) can cause a **peptic ulcer** in the stomach. The excess acid may even make its way into the esophagus, a condition known as **acid reflux**. It causes the condition commonly known as **heartburn**. If left untreated, it can damage the epithelial lining of the esophagus and cause rotting of teeth. Many prescription, as well as over-the-counter, antacids are now available to lower the acidity.

Name: _____ Lab Section: _____

|||||| Review Questions

1. Which enzyme breaks starch into maltose?

   _____

2. Where in the digestive tract is pepsin found?

   _____

3. For which of the enzymes did you test pH sensitivity?

   _____

4. For which of the enzymes did you test temperature sensitivity?

   _____

5. Is emulsification considered "mechanical" digestion?

   _____

6. Phenol Red will turn yellow in the presence of what?

   _____

   _____

7. What happened to amylase when it was boiled?

   _____

   _____

8. Which of the tests revealed the importance of increasing surface area for chemical digestion?

   _____

   _____

9. Discuss how bile and lipase work together to digest fats.

   _____

   _____

10. In the tests for the action of amylase, half of your tubes were controls. Why was this important?

    _____

    _____

Cardiovascular System 1: Anatomy of the Heart and Vessels

objectives

- To become familiar with the anatomy of the heart and blood vessels
- To trace the flow of blood through the heart and entire body
- To distinguish between microscopic specimens of arteries and veins

materials

- human heart models
- latex or rubber gloves
- fetal pigs
- pig or sheep hearts
- dissecting trays
- dissecting instruments
- microscopes
- protective eyewear
- slides of cardiac muscle
- slides of arteries and veins

⚠ SAFETY ALERT!

Preserved specimens contain chemicals that are potentially irritating to the skin and eyes. Do not handle specimens without gloves or eyewear. Dissection instruments are sharp. Take care not to puncture or cut your skin. Handle the microscopes carefully, and always carry them with two hands. Remember that the slides are made of glass. Even when not broken, the edges are sharp. In the event that a slide is broken, inform the instructor and seek guidance concerning cleanup procedures!

▥ Introduction

The **cardiovascular** system includes the **heart, blood vessels,** and **blood.** Together they form a transport system that delivers oxygen and nutrients to and removes waste from all tissues.

In this exercise, you will focus on the anatomy of the heart and vessels and the path of blood flow.

Heart

The **heart** is a hollow ball of cardiac muscle about the size of your fist (**Figure 8-1**). It serves as a pump to ensure nonstop flow of blood through the vessels. Located between the two lungs, the bulk of the heart lies to the left of the sternum.

Anatomically, it is divided into four inner chambers. The two chambers on the top are the **atria** (**singular "atrium"**). The two lower, larger chambers are the **ventricles.** With regards to the beating (pumping) of the heart, the atria will work together, followed by a cooperative effort by the ventricles. With respect to blood flow, however, the heart is actually a double pump with distinct left and right halves.

Vessels

Two major types of blood vessels exist: **arteries** and **veins**. Arteries are vessels that carry blood away from the heart. Veins carry blood to the heart. Blood flows in one direction, from the arteries to the veins. Between the two are very tiny vessels called **capillaries** that have walls that are only one cell thick. Here gas exchange occurs.

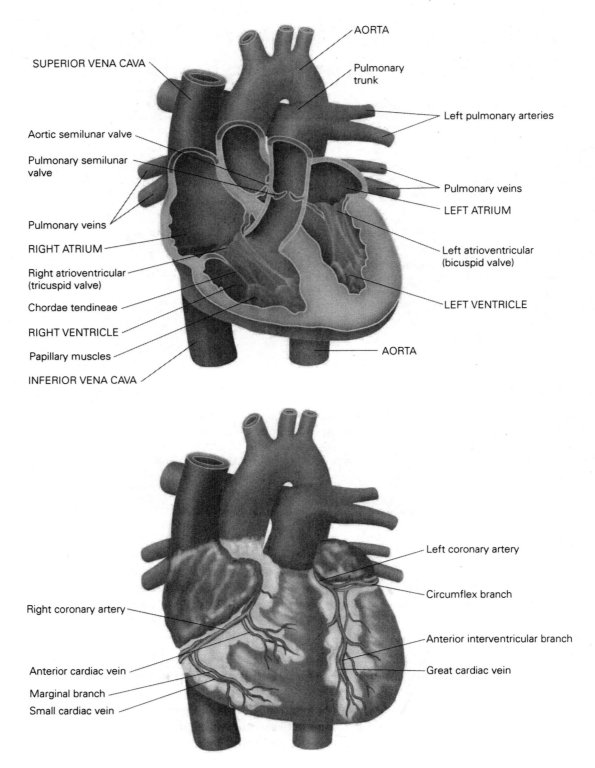

Figure 8-1 Anatomy of the heart.

The differences in the chemical composition of blood and blood pressure in arteries and veins are reflected in their distinct anatomies. Arteries are under higher pressure and have thicker walls, whereas veins have valves to prevent backflow of blood.

procedure

1 Using the human heart models, sheep or pig hearts, and Figure 8-1 locate the following structures:
- right and left atrium
- tricuspid and bicuspid valves
- right and left ventricles
- chordae tendinea
- superior vena cava
- inferior vena cava
- aorta
- pulmonary arteries
- pulmonary trunk
- pulmonary veins
- pulmonary and aortic semi-lunar valves

2 Using Figure 8-2 and Figure 8-3 and any available models, identify the following vessels:

Arteries	Veins
aorta	superior & inferior vena cavae
brachiocephalic trunk	brachiocephalic
subclavian	subclavian
common carotids	internal jugular
brachial	external jugular
celiac trunk	renal
renal	common iliac
common iliac	external iliac
external iliac	hepatic portal
femoral	femoral

3 Flow of blood through the heart, lungs, and body

Blood flows in a continuous one-way conduction system through the entire body. Regarding gas exchange, it is usually divided into **pulmonary** and **systemic** circuits (Figure 8-4). The pulmonary circuit is the flow of blood between the heart and lungs. The systemic circuit is the flow of blood between the heart and the rest of the body tissues.

The human **circulatory system** is a closed system. That is, the only way in or out is by way of diffusion through the walls of the capillaries. Therefore, any place along the route can serve as a starting point to trace the flow through the entire "circle."

- We begin at the **right atrium**. Here, blood low in oxygen is pushed through the **tricuspid valve** into the **right ventricle**.

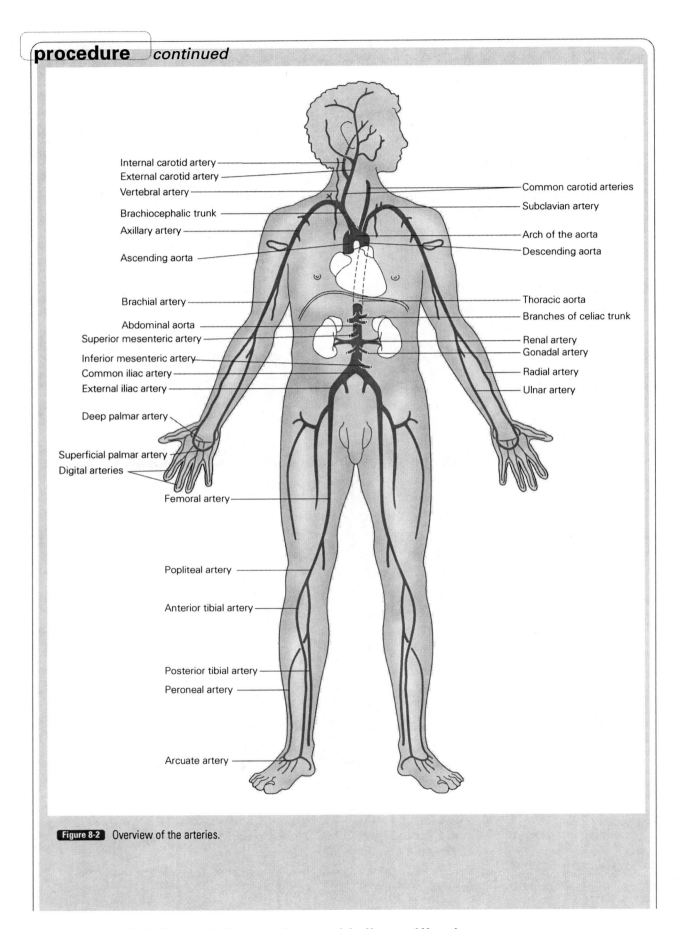

Internal carotid artery
External carotid artery
Vertebral artery
Brachiocephalic trunk
Axillary artery
Ascending aorta
Brachial artery
Abdominal aorta
Superior mesenteric artery
Inferior mesenteric artery
Common iliac artery
External iliac artery
Deep palmar artery
Superficial palmar artery
Digital arteries
Femoral artery
Popliteal artery
Anterior tibial artery
Posterior tibial artery
Peroneal artery
Arcuate artery

Common carotid arteries
Subclavian artery
Arch of the aorta
Descending aorta
Thoracic aorta
Branches of celiac trunk
Renal artery
Gonadal artery
Radial artery
Ulnar artery

Figure 8-2 Overview of the arteries.

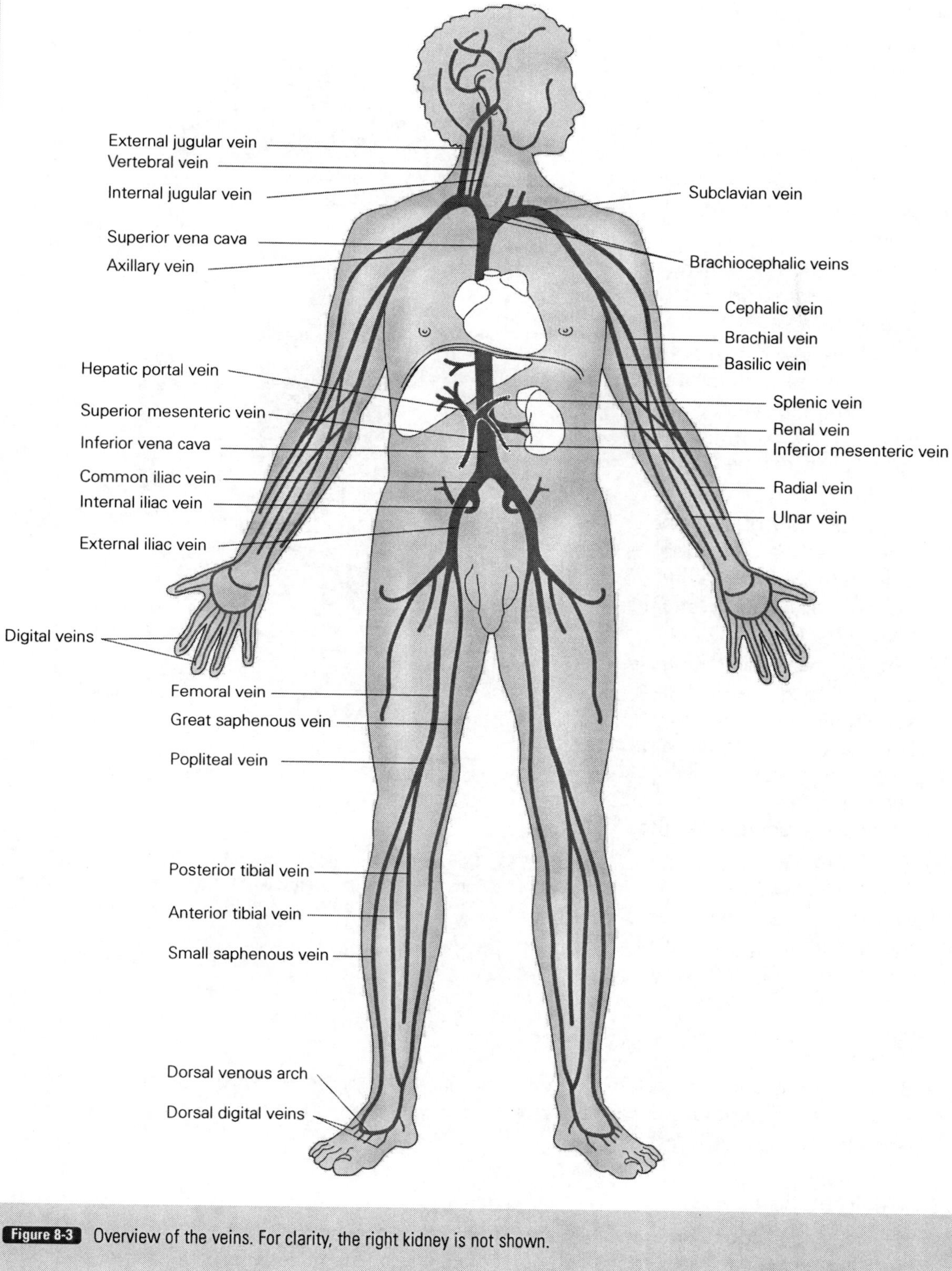
External jugular vein
Vertebral vein
Internal jugular vein
Superior vena cava
Axillary vein
Hepatic portal vein
Superior mesenteric vein
Inferior vena cava
Common iliac vein
Internal iliac vein
External iliac vein
Digital veins
Femoral vein
Great saphenous vein
Popliteal vein
Posterior tibial vein
Anterior tibial vein
Small saphenous vein
Dorsal venous arch
Dorsal digital veins
Subclavian vein
Brachiocephalic veins
Cephalic vein
Brachial vein
Basilic vein
Splenic vein
Renal vein
Inferior mesenteric vein
Radial vein
Ulnar vein

Figure 8-3 Overview of the veins. For clarity, the right kidney is not shown.

- Blood is then sent past the **pulmonary semilunar valve** and through the **pulmonary trunk** into the **pulmonary arteries**. This is the beginning of the pulmonary circuit as blood is about to enter the lungs to engage in gas exchange.
- In the **capillaries** in the lungs, oxygen is picked up, and CO_2 is removed.
- The blood now rich in oxygen returns to the heart by way of the **pulmonary veins.** They transport the blood to the **left atrium.** This completes the pulmonary circuit.
- From the **left atrium,** blood moves past the **bicuspid (mitral) valve** into the **left ventricle.** Now, the systemic circuit now begins.
- Blood leaves the left ventricle, passes through the **aortic semilunar valve** into the **aorta,** and then into the large systemic arteries. Then it enters increasingly smaller arteries as it approaches the capillary beds.
- Again, gases will be exchanged. Oxygen will leave the **capillaries** to enter all of the tissues. CO_2, the waste from cells, will flow in the opposite direction into the capillaries.

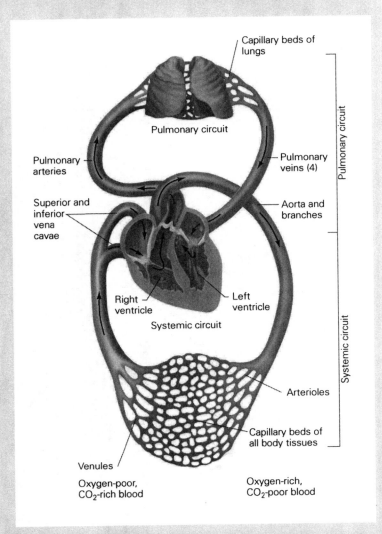

Figure 8-4 The blood pathway includes two circuits. The right ventricle supplies the pulmonary circuit, and the left ventricle supplies the systemic circuit.

- Blood low in oxygen and rich in CO_2 will enter increasingly larger veins until it reaches the **superior** and **inferior vena cavae.** Systemic circulation ends when the vena cavae deliver blood to the **right atrium.**

4 **Fetal circulation**

Fetal circulation is quite different than that after birth. Because the digestive system and lungs are not yet functional, blood gases and nutrients are exchanged between the mother and fetus through the **umbilical veins** and arteries (**Figure 8-5**). The umbilical veins bring blood rich in oxygen and nutrients to the fetus. The **umbilical arteries** carry blood low in oxygen back to

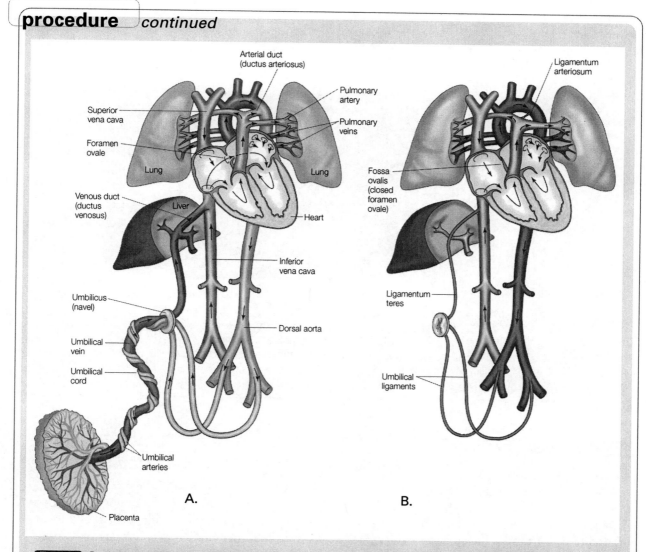

Figure 8-5 Fetal circulation. A. before birth. B. after birth.

the placenta. Most of the blood that flows through the fetus is mixed. That is, it contains varying amounts of oxygen and CO_2. This is sufficient to meet the needs of the fetus.

5 **Dissection**

Using Appendix 2 as your guide, obtain a fetal pig, and begin a dissection of the blood vessels. Note the differences in branching patterns between the human and your specimen.

6 **Slides of arteries, veins, and cardiac muscle**

Using **Figure 8-6** as a guide, differentiate between arteries and veins.

a. Obtain prepared microscope slides of arteries, veins, and cardiac muscle.

b. Observe the slides first on low (scanning) power, and then increase the magnification.

c. Note the difference in the thickness of arteries and veins.

d. Cardiac muscle should be a review from Exercise 5. Recall that it is striated and contains intercalated discs.

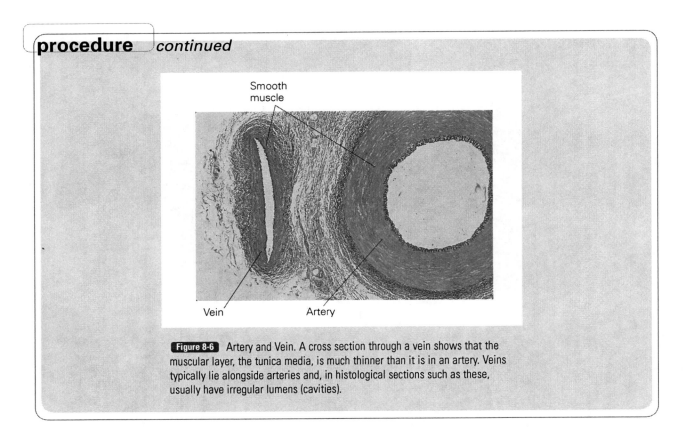

Figure 8-6 Artery and Vein. A cross section through a vein shows that the muscular layer, the tunica media, is much thinner than it is in an artery. Veins typically lie alongside arteries and, in histological sections such as these, usually have irregular lumens (cavities).

CLINICAL CONSIDERATIONS

Atherosclerosis

Atherosclerosis is a disease caused by "plaques" in the walls of arteries. The liver and small intestines produce molecules called lipoproteins. They are a unique combination of lipids, phospholipids, proteins, and cholesterol. There are two types: **high-density lipoproteins (HDLs)** and **low-density lipoproteins (LDLs)**. LDLs carry cholesterol to cells for maintenance and repair; however, too many LDLs can lead to a buildup of plaques of cholesterol, or atherosclerosis. This can lead to reduced blood flow and possibly cardiac problems, such as a heart attack. HDLs, or the "good cholesterol," on the other hand, deliver excess cholesterol to the liver for elimination.

Name: _____ Lab Section: _____

IIIIII Review Questions

1. Which chambers of the heart have the thickest walls?

   _____

   _____

2. Is the blood found in the left atrium rich in oxygen?

   _____

   _____

3. Which valve prevents the backflow of blood into the right atrium?

   _____

   _____

4. Which vessels return blood from the lungs to the heart?

   _____

   _____

5. Is the inferior vena cava considered a structure in pulmonary circulation?

   _____

   _____

6. Which have thicker walls, arteries or veins?

   _____

   _____

7. Blood that passes by the tricuspid valve will be rich in which gas?

   _____

   _____

8. Which artery supplies the kidney with oxygen rich blood?

   _____

   _____

9. Which vessels are responsible for gas exchange?

   _____

   _____

Name: _____ Lab Section: _____

10. List all of the heart valves and where they are located.

_____

_____

_____

_____

_____

_____

11. Starting at the right atrium, list the structures through which the blood will pass if it makes a complete cycle back to the right atrium.

_____

_____

_____

_____

_____

_____

_____

Name: _____ Lab Section: _____

12. Label the diagrams of blood vessels.

Arteries

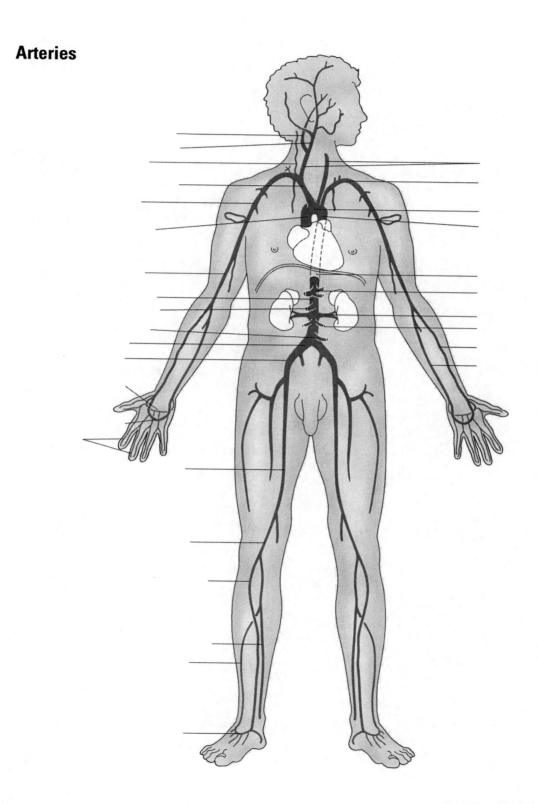

Name: _____ Lab Section: _____

Veins

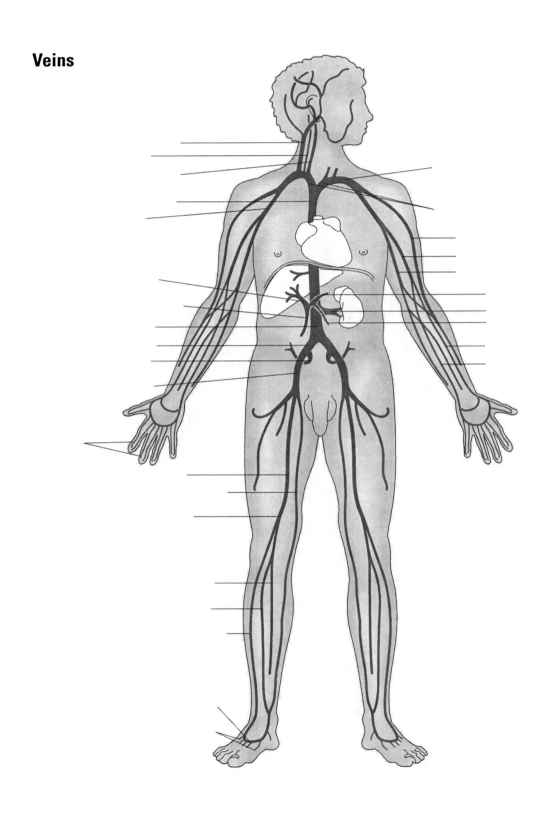

Cardiovascular System 2: Functions of the Blood, Heart, and Vessels

objectives

- To identify red and white blood cells and platelets under the microscope
- To identify the regions on an EKG
- To measure heart rate and blood pressure during rest and exercise

materials

- slides of blood
- microscopes
- blood pressure cuffs
- stethoscopes
- exercise step

SAFETY ALERT!

Handle the microscopes carefully, and always carry them with two hands. Remember that the slides are made of glass. Even when not broken, the edges are sharp. In the event that a slide is broken, inform the instructor and seek guidance concerning cleanup procedures! When using human subjects, take care to treat them respectfully and not to cause them any discomfort.

|||||| Introduction

Now that you have a solid foundation in cardiovascular anatomy, you can investigate the functions of blood, the heart, and vessels.

In this exercise, you will observe slides of blood and use human subjects to investigate heart rate and blood pressure under varying conditions.

procedure

1 Blood

As was discussed in Exercise 5, blood is comprised of the fluid called plasma and the formed elements: red blood cells, white blood cells, and platelets (**Figure 9-1**). Plasma is mostly water but also contains proteins and ions such as sodium, potassium, and calcium.

Red blood cells, or **erythrocytes**, are small cells that appear as little discs and have no nucleus. They are filled with the red pigment hemoglobin that carries oxygen. They make up the majority of blood cells.

White blood cells, or **leukocytes**, are much larger than red blood cells. They have nuclei and are involved in immune responses and fighting infections. There are five different types of white blood cells, all with different functions (**TABLE 9-1**).

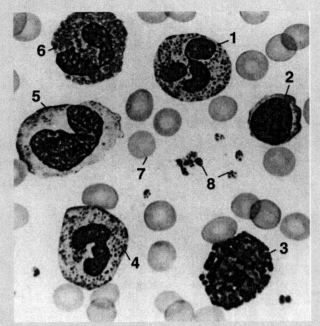

Figure 9-1 Human blood, showing formed elements (1000×).
1. neutrophil
2. lymphocyte
3. basophyl
4. band cell (immature neutorphil)
5. monocyte
6. eosinophyl
7. erythrocyte
8. platelets

Platelets are only fragments of cells and are very small even compared with red blood cells. They function in clotting in order to prevent extensive blood loss.

Using Figure 9-1 and Table 9-1 as guides, identify red blood cells, the various white blood cells, and platelets under the microscope.

a. As always, begin on low power.

b. Switch to higher power carefully. Here is where you should be able to distinguish between red blood cells, the various white blood cells, and platelets.

2 Heart beat and heart rate

The heart has its own specialized conduction system to ensure continuous rhythms, or heart beats. The **SA (sinoatrial) node**, often referred to as the **pacemaker**, is located in the upper region of the right atrium (**Figure 9-2**). As the name implies, it is responsible for setting the "pace" of the heart, or heart rate. Heart rate is usually measured in beats per minute.

When stimulated by the nervous system, the SA node sends signals to the atria, causing them to "contract" and pump blood into the ventricles.

At the same time, signals are sent to the **AV (atrioventricular) node** lying between the right atrium and right ventricle. The AV node sends signals down the **AV bundle** that terminate in the **Purkinje fibers.** These communicate with the cardiac muscle in the walls of the ventricles. The ventricles then contract causing blood to leave the heart.

This is depicted on an **EKG (electrocardiogram)**: a graphic representation of the electrical activity of cardiac muscle during a heart beat, or **cardiac cycle** (Figure 9-2).

An EKG has several notable regions. The **P wave** represents contraction of the atria, and the **QRS wave** represents ventricular contraction. The **T wave** depicts the electrical activity that occurs while the cardiac muscle in all four chambers is relaxing and awaiting another stimulus from the SA node.

An EKG can be used as a diagnostic tool to identify possible problems with the heart.

TABLE 9-1	Summary of Blood Cells			

Name	Description	Concentration (Number of Cells/mm³)	Life Span	Function
Red blood cells (RBCs)	Biconcave disk; no nucleus	4 to 6 million	120 days	Transports oxygen and carbon dioxide
White blood cells (WBC) Neutrophil	Approximately twice the size of RBCs; multilobed nucleus; clear-staining cytoplasm	3000 to 7000	6 hours to a few days	Phagocytizes bacteria
Eosinophil	Approximately same size as neutrophil; large pink-staining granules; bilobed nucleus	100 to 400	8 to 12 days	Phagocytizes antigen–antibody complex; attacks parasites
Basophil	Slightly smaller than neutrophil; contains large, purple cytoplasmic granules; bilobed nucleus	20 to 50	Few hours to a few days. . .	Releases histamine during inflammation
Monocyte	Larger than neutrophil; cytoplasm grayish blue; no cytoplasmic granules; U- or kidney-shaped nucleus	100 to 700	Lasts many months	Phagocytizes bacteria, dead cells, and cellular debris
Lymphocyte	Slightly smaller than neutrophil; large, relatively round nucleus that fills the cell	1500 to 3000	Can persist many years	Involved in immune protection, either attacking cells directly or producing antibodies
Platelets	Fragments of megakaryocytes; appear as small dark-staining granules	250,000	5 to 10 days	Play several key roles in blood clotting

3 Measuring heart rate

There are several methods to measure heart rate.

a. Using a stethoscope, listen for the sounds of a heart beat on your lab partner. Place the stethoscope slightly to the left of the middle of the sternum. The two sounds of the heart beat, "lubb-dupp," are caused by the closing of the valves. The first sound is the closing of the AV (bicuspid and tricuspid valves), and the second sound is the closing of the semilunar valves. The two sounds together represent a complete cardiac cycle or heart beat. Therefore, you can use them to measure heart rate. Listen for 1 minute, and record the results.

_____

b. To save time, you can also use the same procedure while listening for only 15 seconds. You then multiply the result by four to get the beats per minute.

_____

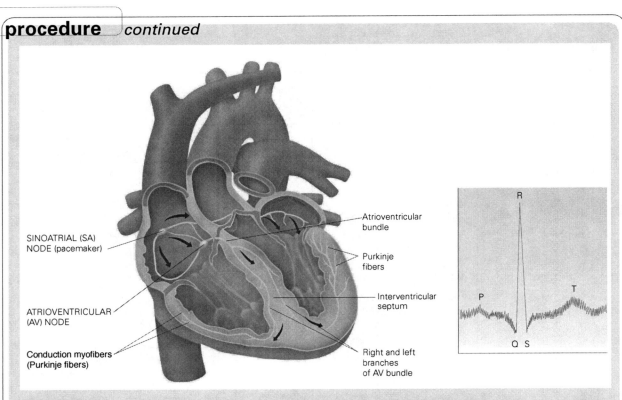

Figure 9-2 The impulse generation and conduction system of your heart. Also shown is the tracing of an EKG. The P wave corresponds to atrial depolarization, the QRS complex to ventricular depolarization, and the T wave to ventricular repolarization.

In medicine, your heart rate is often referred to as your pulse. It can be measured by feeling arteries in the wrist or neck (**Figure 9-3**).

a. Take your partner's resting pulse on various body regions.

b. Record the pulse rate, and note any differences in the "strength" of the pulse in the different vessels.

4 Blood pressure

Clinical blood pressure refers to the pressure exerted on the walls of the large arteries close to the aorta during contraction and relaxation of the left ventricle. It is numerically expressed as the pressure during contraction over the pressure during relaxation. For example, **120/80** is the average blood pressure for a healthy young adult. The top number is known as the **systolic** (state of contraction) and the bottom as the **diastolic** (state of relaxation). Therefore, blood pressure is recorded as systolic/diastolic.

Using a sphygmomanometer, or blood pressure cuff, and a stethoscope, you will measure the blood pressure of your lab partner. Have your partner quietly sit in a chair during this procedure.

a. Place the head of the stethoscope on the inside of the elbow joint on the left arm (**Figure 9-4**). This will be in the region of the brachial artery. Place the other end into your ears. Wrap the blood pressure cuff around the arm, about 2 inches above the stethoscope.

b. Pump the bulb until it builds up pressure enough to go to a reading of about 180 mm Hg.

c. Slowly release the valve while watching the needle on the meter. At some point, the pressure of the cuff will match the pressure inside the artery. At this point, you will begin to hear the pulse. The reading on the meter at this point will be the systolic, or top number.

d. Continue to release the pressure while watching the needle on the meter. At some point, the pulse sounds will disappear. The reading on the meter at this point will be the diastolic, or bottom number.

e. Record the results:

_____

_____

_____

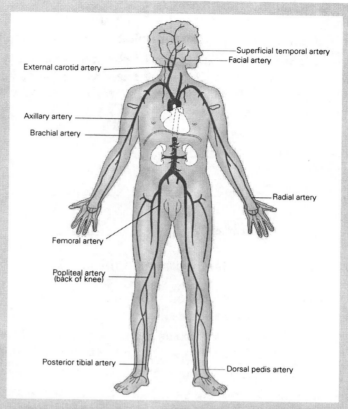

Superficial temporal artery
Facial artery
External carotid artery
Axillary artery
Brachial artery
Radial artery
Femoral artery
Popliteal artery (back of knee)
Posterior tibial artery
Dorsal pedis artery

Figure 9-3 Arteries to measure pulse.

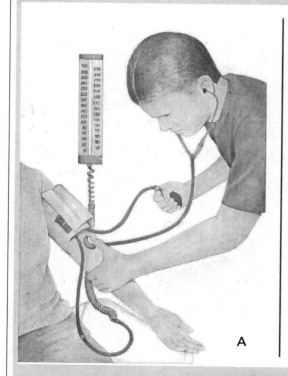

A

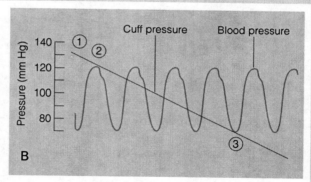

Cuff pressure   Blood pressure

Pressure (mm Hg)

140 — ①②
120 —
100 —
80 —

③

B

1. No sound is heard. Cuff pressure is gradually released. When pressure in the cuff falls below the arterial pressure, blood starts flowing through the artery once again. This is the systolic pressure.

2. The first sound will be heard. Cuff pressure continues to drop. When cuff pressure is equal to the lowest pressure in the artery, the artery is fully open and no sound is heard. This is the diastolic pressure.

Figure 9-4 Blood Pressure Reading (A) A sphygmomanometer (blood pressure cuff) is used to determine blood pressure. (B) As shown, the blood pressure rises and falls with each contraction of the heart. When the pressure in the cuff exceeds the arterial peak pressure, blood flow stops.

Exercise 9: Cardiovascular System 2: Functions of the Blood, Heart, and Vessels

5 **Measuring heart rate and blood pressure under varying conditions**

a. Measure heart rate and blood pressure while your partner is standing and after light and medium exercise. Light exercise could be walking at a slightly faster than normal pace for two minutes. Medium exercise could be jogging or running in place for 1 to 2 minutes. Record the results.

Standing _____

Light exercise _____

Medium exercise _____

6 **Harvard Step Test: personal fitness test**

This procedure allows you to test the cardiovascular "fitness" of your lab partner using heart rate after exercise as an indicator.

a. Using a 16- to 20-inch step, the subject will step up and back down. The entire cycle of up and down should take 2 seconds. Therefore, 30 cycles should be completed in 1 minute.

b. The subject will do this for at least 3, but no more than 5, minutes.

c. After the exercise, the subject will rest for 1 minute. At the end of this minute, the subject's 30 second pulse should be recorded. Record 30 second pulse also after 2 minutes and after 3 minutes.

30 second pulse, 1 minute after exercise _____

30 second pulse, 2 minutes after exercise _____

30 second pulse, 3 minutes after exercise _____

d. The Personal Fitness Index (PFI) can then be calculated using the following formula:

PFI = number of seconds of exercise × 100/2 (sum of all pulse counts)

For example, if pulse counts were 52, 50, and 49 after 3 minutes of exercise (180 seconds), the PFI would be as follows:

180 × 100/2 (52 + 50 + 49) = 59.6

e. Using this scale, determine fitness level.
- Below 55: poor condition
- 55 to 64: low average condition
- 65 to 79: high average condition
- 80 to 90: good condition
- Over 90: excellent condition

Blood Pressure

High blood pressure, or **hypertension**, is defined as resting blood pressure that is persistently above 140/90. Hypertension is the most common cardiovascular imbalance. Although many suffering from hypertension are genetically predisposed to such a condition, other contributing factors include stress, diet, and lack of exercise. Hypertension has been called the "silent killer" because symptoms are often not detected until it is almost too late for treatment. Left untreated, it can cause heart and kidney failure and strokes.

Blood pressure ranges are as follows:
Optimal: 120/80
Normal range: systolic 120 to 139, diastolic 85 to 89
Hypertension: systolic 140 and above, diastolic 90 and above

Medications are available to control blood pressure, but lifestyle changes can have a dramatic effect. These include exercise, reduced sodium intake, stress management, and limited alcohol consumption.

Name: _____ Lab Section: _____

||||||| Review Questions

1. What does the P wave represent?

   _____

   _____

2. Which valves, closing make up the second sound of a heart beat?

   _____

   _____

3. Which white blood cells are the largest?

   _____

   _____

4. What is the formal name for red blood cells?

   _____

   _____

5. Which of the formed elements aids in clotting?

   _____

   _____

6. What does the T wave represent?

   _____

   _____

7. Which chambers show a stronger electrical output during a cardiac cycle?

   _____

   _____

8. What is meant by "systolic?"

   _____

   _____

Name: _____ Lab Section: _____

9. While measuring blood pressure, which of the numbers do you record when you no longer hear the pulse pressure?

_____

_____

10. List three places that you can use to measure pulse. How do they differ with respect to the pressure you feel? Why?

_____

_____

_____

11. In general, how were blood pressure and heart rate affected by exercise? According to the results of the Harvard Step Test, are the students "fit?"

_____

_____

Respiratory System

objectives

- To become familiar with the anatomy of the respiratory system
- To measure lung volumes and capacities

materials

- human respiratory system models
- fetal pigs
- dissecting trays
- dissecting instruments
- protective eyewear
- wet spirometer
- latex or rubber gloves

⚠ SAFETY ALERT!

Preserved specimens contain chemicals that are potentially irritating to the skin and eyes. Do not handle specimens without gloves or protective eyewear. Dissection instruments are sharp. Take care not to puncture or cut your skin! Proceed with caution and respect when using human subjects.

||||||| Introduction

Respiration is the process by which energy is built for metabolic needs. Most organisms, including humans, build energy by consuming oxygen. Our respiratory system, therefore, is designed to capture and transport oxygen.

The two most prominent organs in the respiratory tract are clearly the **lungs** (**Figure 10-1**). They are comprised of soft, spongy tissue that is actually a swelling of the end of the respiratory tract. The tract includes the **oral** and **nasal cavities, pharynx, larynx, epiglottis** (**Figure 10-2**), **trachea, bronchial tubes,** and the tiny **alveolar sacs** where gas exchange occurs. The bronchial tubes and the alveoli are actually part of the lungs.

Breathing, or **ventilation,** is the act of moving air into and out of the lungs. **Inhalation,** also called **inspiration,** brings air rich in oxygen into the lungs. The oxygen diffuses into the capillaries surrounding the alveoli. At the same time, CO_2, the waste product from metabolism, leaves the capillaries and enters the alveoli. The air now filled with CO_2 is then moved out. This is called **exhalation,** or **expiration.**

In this exercise, you will study the structure of the respiratory tract and measure lung volumes and capacities of your lab partners.

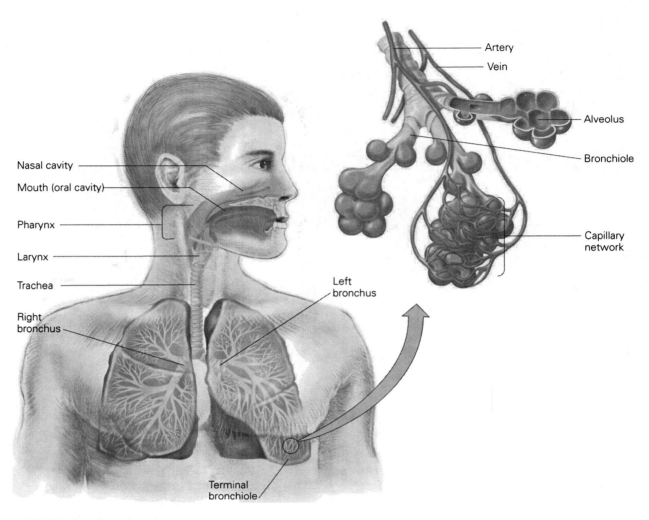

Nasal cavity

Mouth (oral cavity)

Pharynx

Larynx

Trachea

Right bronchus

Left bronchus

Terminal bronchiole

Artery

Vein

Alveolus

Bronchiole

Capillary network

Figure 10-1 Overview of respiratory tract

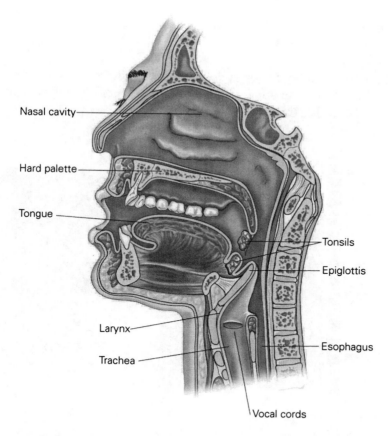

Figure 10-2 Upper portion of respiratory tract showing the relationship between the epiglottis, trachea, and esophagus.

Labels in figure:
- Nasal cavity
- Hard palette
- Tongue
- Tonsils
- Epiglottis
- Larynx
- Esophagus
- Trachea
- Vocal cords

procedure

Respiratory Anatomy and Pathway for Air Transport

Using Figures 10-1 and 10-2, Appendix 2, human respiratory models, and fetal pigs, locate the following structures. Structures are listed in the order in which air passes during inhalation. During exhalation air will travel in reverse.

Oral cavity (mouth): air taken in

Nasal cavity: air taken in

Pharynx: back of oral cavity

Epiglottis: flap of connective tissue at the top of the larynx that ensures that air, not food or liquid, enters the trachea

Larynx (voice box): needs air to produce sound for speech

Trachea (windpipe): directs air into the bronchial tubes

Bronchial tubes: branch off from the trachea and enter both lungs

Alveolar sacs: the end of the passageway for air; the terminal ends of the bronchial tubes; surrounded by capillaries; place of gas exchange

Lungs: two organs (right and left) made of soft, spongy tissue that house the bronchial tubes and alveolar sacs

Lung Volumes and Capacities

The amount of air that passes through the lungs during normal and exaggerated breathing has several components that can be measured. These are referred to as **lung volumes** and **capacities** and are listed below and in **Figure 10-3** .

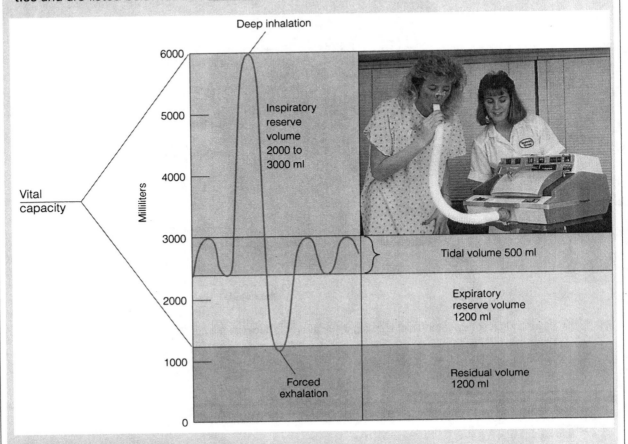

Figure 10-3 Graph of lung volumes and capacities

Tidal volume (TV): the amount of air that passes through the lungs during a normal respiratory cycle of inspiration followed by expiration

Inspiratory reserve volume (IRV): the amount of air that can be inspired above and beyond TV

Expiratory reserve volume (ERV): the amount of air that can be expired above and beyond TV

Vital capacity (VC): the amount of air that can be inspired and expired above and beyond TV; the maximum volume that can be used at any time

Residual volume (RV): the amount of air that is never exchanged; ensures that there is always some air in the respiratory tract so it does not collapse

Measurement of Lung Volumes and Capacities

Using a wet spirometer, you will measure and calculate lung volumes and capacities. This is accomplished by displacing a volume of water measured in milliliters. Your lab instructor will calibrate the spirometer.

1 Tidal Volume

a. To measure TV, maintain a normal breathing rhythm for 10 seconds.

b. After one of your inspirations, put the mouthpiece into your mouth while holding your nose closed so that no airs escapes through the nasal cavity. Exhale normally as you have done for the past 10 seconds. The amount of water displaced will be the TV. Repeat three times, and record the average TV:

1st TV _____

2nd TV _____

3rd TV _____

average TV = 1st + 2nd + 3rd/3 _____

2 ERV

a. Maintain a normal breathing rhythm for 10 seconds.

b. After one of your inspirations, exhale (expire) as hard and long as you can through your mouth while holding your nose closed. The amount of water displaced will be your ERV. Record the results.

ERV _____

3 VC

a. Maintain a normal breathing rhythm for 10 seconds.

b. Inhale as long and hard as possible. Then put the mouthpiece in your mouth and exhale as long and hard as possible while holding your nose closed. The amount of water displaced will be your VC.

c. Record the results

VC _____

4 IRV

For sanitary reasons and to avoid health risks, you do not want to inhale through the spirometer tube. Therefore, you will measure IRV indirectly based on the following calculations.

$$VC = IRV + ERV$$

Because you have already measured VC and ERV, you can now use this equation:

$$IRV = VC - ERV$$

Calculated IRV _____

5 Minute Ventilation

The minute ventilation is the amount of air inhaled and exhaled in 1 minute. It is calculated by multiplying **breathing rate (breaths per minute)** by TV:

$$\text{breathing rate} \times \text{TV} = \text{minute ventilation}$$

You have already calculated TV. Now measure breathing rate and record:

Breathing rate _____

Calculate MV and record.

MV _____

CLINICAL CONSIDERATIONS

Asthma

Asthma is actually a combination of respiratory disorders that together obstruct the passage of air. It usually results from allergic reactions to dust, animal dander, pollen, and molds. Even certain foods can bring about these allergic reactions. Emotional distress, exercise, and cigarette smoke can also trigger asthmatic attacks.

The allergic reaction causes an increase in mucus production and the constriction of the bronchiole tubes. Together, these greatly reduce the flow of air.

Chronic inflammation of the epithelial lining of the respiratory tract due to allergic reactions also can contribute to the severity of asthma.

Symptoms include coughing, difficulty breathing, wheezing, and possibly anxiety, a generic disorder that alerts the body that physiologic problems are present. In this case, there is a lack of oxygen.

Treatment may involve a combination of an inhalant that contains epinephrine to open bronchiole tubes and anti-inflammatory drugs.

Those suffering from asthma are usually children who often exhibit less severe symptoms as they grow older.

Name: _____ Lab Section: _____

||||||| Review Questions

1. Where does gas exchange occur in the respiratory tract?

    _____

    _____

2. Is exhaled air rich in oxygen?

    _____

    _____

3. The voice box is technically called what?

    _____

    _____

4. Which lung capacity measures the amount that can be inhaled and exhaled in one huge breath?

    _____

    _____

5. VC minus IRV will equal what?

    _____

    _____

6. The lungs house the bronchial tubes and which other structures?

    _____

    _____

7. Which structure prevents food and liquid from entering the trachea?

    _____

    _____

8. The back of the oral cavity is called what?

    _____

    _____

Name: _____   Lab Section: _____

9. Trace the flow of air during exhalation.

_____

_____

10. Label the diagram.

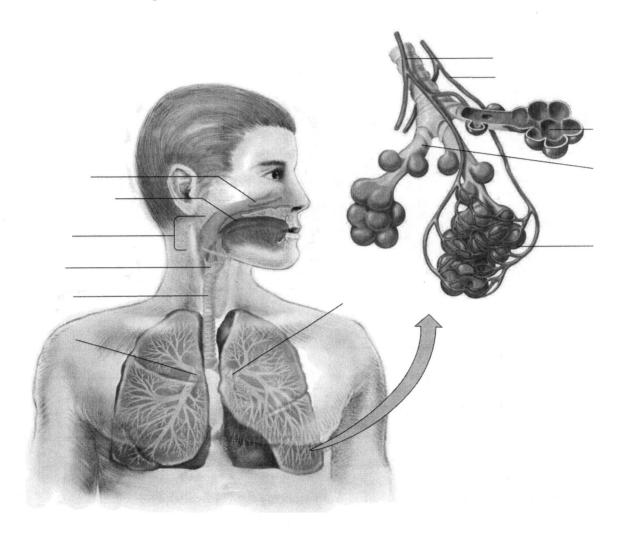

Skeletal System

objectives

- To identify the regions of a long bone
- To locate and identify the microscopic structures in a bone slide
- To locate and identify the two divisions of the skeleton and all of the bones and selected landmarks
- To determine age and gender of skeletal remains

materials

- long bone
- bone slides
- microscopes
- real or plastic articulated and disarticulated skeletons
- atlas (if available)

⚠ SAFETY ALERT!

Carry the microscopes with two hands. The slides are sharp even if not broken. In the event that a slide is broken, consult your instructor concerning cleanup procedures. Handle human skeletal remains with care and respect.

⫼⫼ Introduction

The skeletal system consists of bones, cartilage, tendons, ligaments, and joints. In this lab, we focus exclusively on some of the **206 bones** that make up the human skeleton. The human skeleton can be divided into two components: the **axial** skeleton and the **appendicular** skeleton (**Figure 11-1**). The axial skeleton includes the skull, vertebral column, and the ribcage. It is named axial because it makes up part of the center axis of the body. The appendicular skeleton gets its name from the term **appendages**, which refers to the arms and legs. It also includes the bones of the shoulders, hips, hands, and feet.

Bone, or **osseus tissue**, is quite dynamic. That is, it requires nutrients and gives off waste just as any other tissue in the body. Do not be fooled by the dry, hard, lifeless nature of the bones you will observe in the lab.

Osteocytes ("osteo" = bone; "cyte" = cell) produce the hard, calcified matrix found in bone. They are organized in a circular fashion in an **osteon** or **haversian system** (**Figure 11-2**).

Support, protection, and producing movement are the obvious functions of the skeleton. However, bone also serves as a storehouse for minerals such as calcium and phosphorus. Bone **marrow** plays a major role in the formation of blood (**Figure 11-3**).

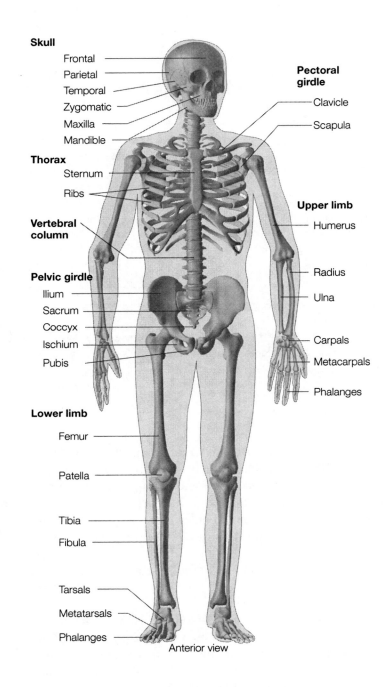

Skull
Frontal
Parietal
Temporal
Zygomatic
Maxilla
Mandible

Thorax
Sternum
Ribs

Vertebral column

Pelvic girdle
Ilium
Sacrum
Coccyx
Ischium
Pubis

Lower limb
Femur
Patella
Tibia
Fibula
Tarsals
Metatarsals
Phalanges

Pectoral girdle
Clavicle
Scapula

Upper limb
Humerus
Radius
Ulna
Carpals
Metacarpals
Phalanges

Anterior view

Figure 11-1 The Human Skeleton.

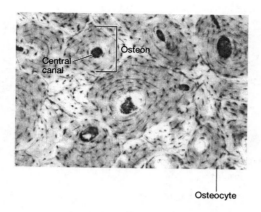

Central canal

Osteon

Osteocyte

Figure 11-2 Microscopic view of compact bone.

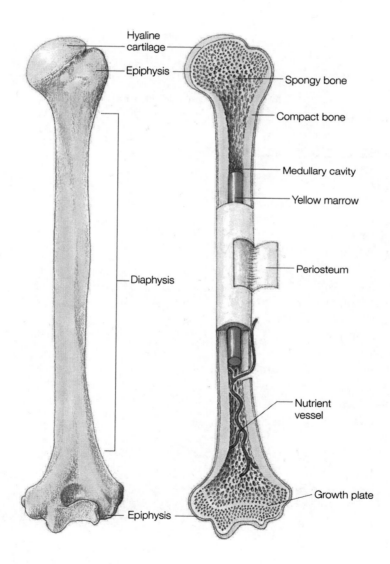

Hyaline cartilage

Epiphysis

Spongy bone

Compact bone

Medullary cavity

Yellow marrow

Diaphysis

Periosteum

Nutrient vessel

Epiphysis

Growth plate

Figure 11-3 Anatomy of Long Bones.

procedure

1 **Observation of a long bone and microscope structures.**

Using your textbook and Figures 11-2 and 11-3, identify the following structures:

Long bone

- epiphysis
- medullary cavity, where marrow is found
- diaphysis
- growth plates, where epiphysis meets diaphysis
- periosteum—connective tissue, will not be found on dry bones

Compact bone slide (Observe under low and high magnification)

- haversian canal
- osteocytes
- canaliculi
- matrix

2 **Skeleton**

Using the lab manual, textbook, and an atlas if available, identify all of the bones and selected landmarks listed here.

Axial Skeleton

    Skull (Figure 11-4A   Figure 11-4B   Figure 11-4C)

        parietal bones

        frontal bone

        nasal bones

        mandible

        zygomatic bone

        occipital bone: foramen magnum

        maxilla

        temporal bone: mastoid process

        sutures: nonmovable joints between bones in the skull

        sagittal, lambdoid, squamosal, and coronal

    Vertebral column (Figure 11-5)

        cervical vertebrae and curvature

        thoracic vertebrae and curvature

        lumbar vertebrae and curvature

        sacrum and coccyx

    Ribcage (Figure 11-1)

        12 pairs of ribs

        sternum

Appendicular Skeleton

    Shoulders, arms, and hands (Figure 11-1)

        shoulder or pectoral girdle: scapula and clavicle

        arm: humerus, radius, and ulna

hand: carpals, metacarpals, and phalanges
Hip, legs, and feet (Figure 11-1)
hip or coxal bones: ilium, ischium, and pubis; pubic arch and acetabulum
leg: femur, tibia, and fibula
feet: tarsals, metatarsals, and phalanges
patella: kneecap
Male and female hips

What Can the Bones Tell Us? Age and Gender

Skeletal remains hold a great deal of information that can be used by forensic scientists (crime scene investigators) as well as archeologists and paleontologists. A trained "osteologist," bone scientist, can examine bones to make estimates of age at the time of death and to determine gender. Certain diseases such as arthritis and some bacterial infections, leave a lasting "impression" on bones. Arthritis can cause growth of extra bits of bone in areas that have experienced a large amount of wear. This is known as "lipping." It is usually found in older individuals and can therefore serve as an estimate of age. Males and females differ with regards to the size and shape of certain bones such as those of the skull and hip. These regions are both very reliable sources of gender determination.

Using this list, attempt to determine the gender and approximate age at the time of death of the skeleton in your lab.

Age

1 Growth plates still visible on long bones such as the femur and humerus indicate an age of not more than 25 to 30 years.

2 Complete fusion of the three bones of the hip (ilium, ischium, and pubis) at the acetabulum indicates an age of at least 20 to 22 years.

3 Sutures that are beginning to disappear because of fusion indicate an age of at least 50 years.

4 If the teeth are still intact, check for wear. This is hard to quantify but can at least tell you whether they were very young or very old.

5 Check for lipping of bone on the vertebrae, this usually begins at about age 40 and steadily progresses.

Gender

1 Males will, in general, be larger than females; however, you only need to look around campus to see that this is only a generalization and cannot be used as the only indicator.

2 A male skull will exhibit a very large and roughened mastoid process.

3 Female skulls have very bulbous parietal bones. This is known as the "parietal swelling." Feel both male and female skulls of your classmates. The difference is striking.

4 The chin (front of the mandible) will be somewhat triangular in females and squarer in males.

5 Male hips are high and narrow. Female hips are broader and shorter with a pubic arch angle of greater than 90 degrees. This is an adaptation for childbirth.

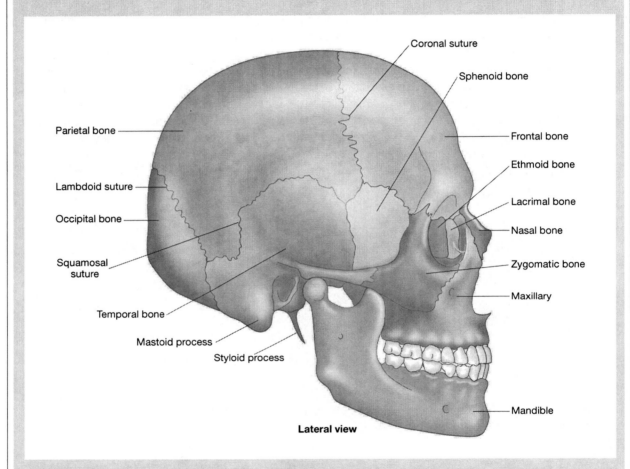

Coronal suture

Sphenoid bone

Parietal bone

Frontal bone

Ethmoid bone

Lambdoid suture

Lacrimal bone

Occipital bone

Nasal bone

Zygomatic bone

Squamosal suture

Maxillary

Temporal bone

Mastoid process

Styloid process

Mandible

Lateral view

Figure 11-4A Bones of the skull. Lateral view.

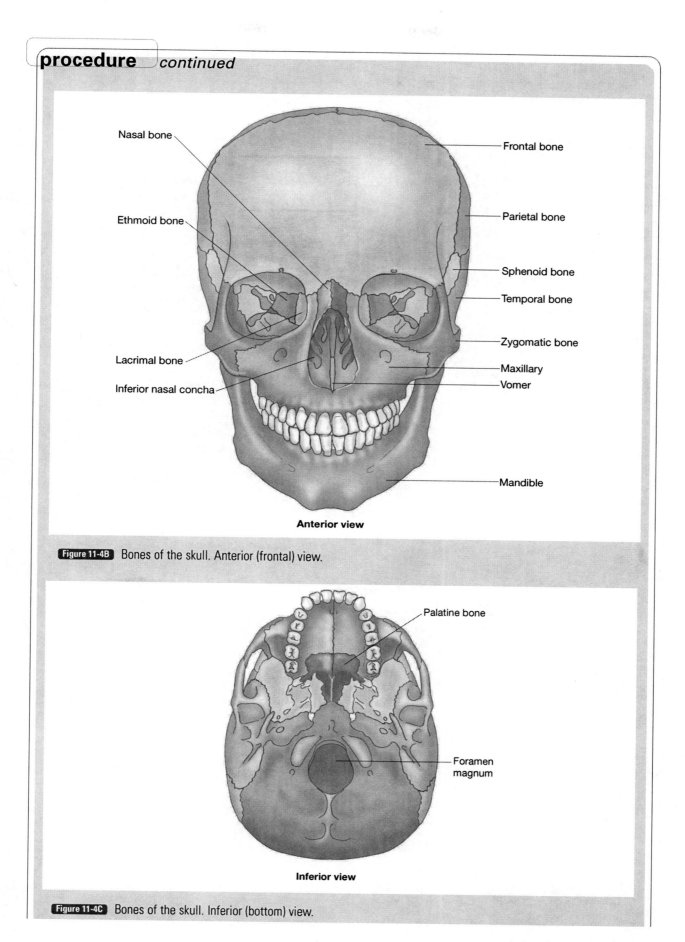

Nasal bone

Ethmoid bone

Lacrimal bone

Inferior nasal concha

Frontal bone

Parietal bone

Sphenoid bone

Temporal bone

Zygomatic bone

Maxillary

Vomer

Mandible

Anterior view

Figure 11-4B Bones of the skull. Anterior (frontal) view.

Palatine bone

Foramen magnum

Inferior view

Figure 11-4C Bones of the skull. Inferior (bottom) view.

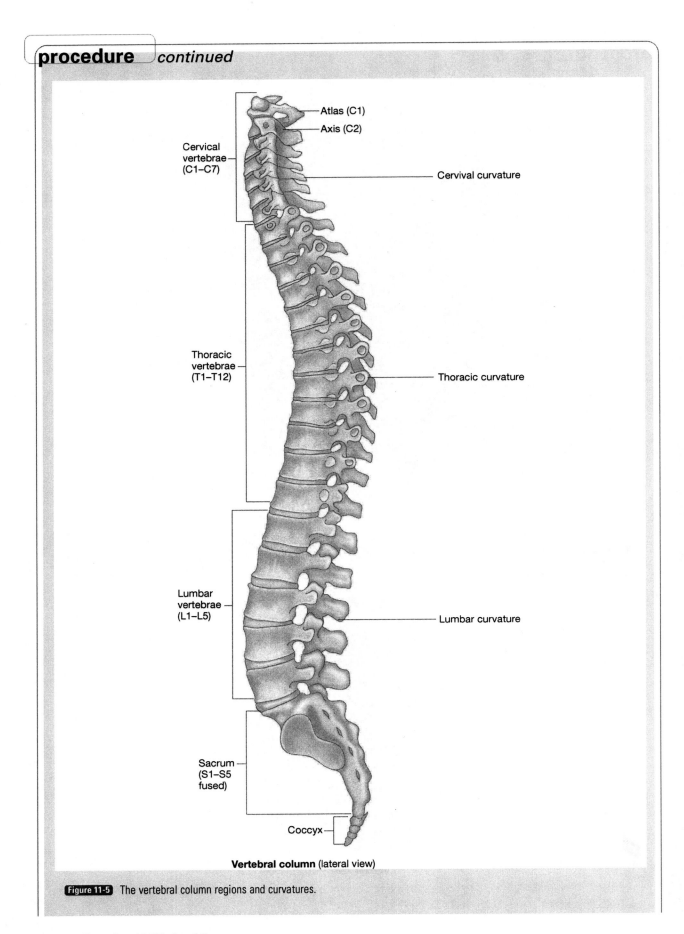

Atlas (C1)

Axis (C2)

Cervical
vertebrae
(C1–C7)

Cervival curvature

Thoracic
vertebrae
(T1–T12)

Thoracic curvature

Lumbar
vertebrae
(L1–L5)

Lumbar curvature

Sacrum
(S1–S5
fused)

Coccyx

Vertebral column (lateral view)

Figure 11-5 The vertebral column regions and curvatures.

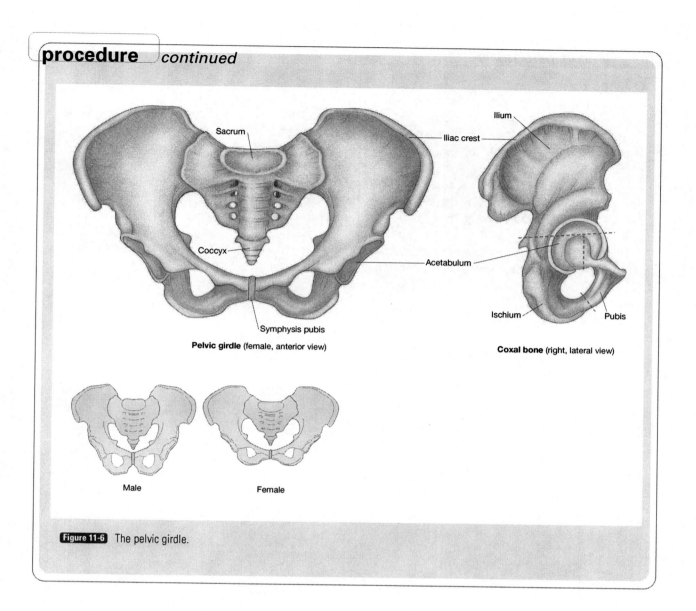

Sacrum

Iliac crest

Ilium

Coccyx

Acetabulum

Ischium

Pubis

Symphysis pubis

Pelvic girdle (female, anterior view)

Coxal bone (right, lateral view)

Male

Female

Figure 11-6 The pelvic girdle.

Osteoporosis

Osteoporosis literally means "porous bones." This condition occurs when calcium loss exceeds resorption. This is a natural phenomenon as we age because of decreasing levels of estrogen and testosterone in women and men, respectively. These hormones both promote the growth of bone. Women are more susceptible, especially after menopause, when estrogen production drops dramatically.

The decrease in bone density leads to the inability of bones to resist stress and forces in everyday movements. This results in fractures that are most often in the hips, vertebrae, and hands. Osteoporosis can cause height loss because of the shrinkage of bones and can also cause associated pain.

Treatment includes increasing dietary calcium and vitamin D. In postmenopausal women, treatment may also include estrogen replacement therapy.

Name: _____   Lab Section: _____

|||||| Review Questions

1. Which bones are closer to the scapula, the carpals or metacarpals?

   _____

2. What are the small passageways that connect the haversian canal to the osteocytes?

   _____

3. How many pairs of ribs do humans have?

   _____

4. Which is not part of the appendicular skeleton?
   a. sternum
   b. tarsals
   c. carpals
   d. humerus

   _____

5. List all of the functions of the skeletal system.

   _____

   _____

   _____

   _____

   _____

   _____

6. What was the estimate of age for the skeleton in your lab? What gender did you assign it? Was there any conflicting information? If so, what?

   _____

   _____

   _____

   _____

Name: _____ Lab Section: _____

7. Label the diagram of the skeleton.

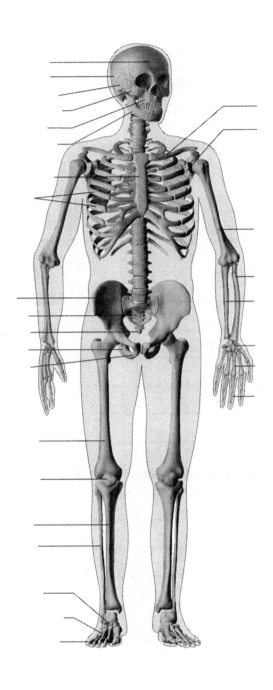

Skeletal Musculature

- To identify select skeletal muscles on models of humans and dissected specimens.
- To gain an understanding of how muscles are named and classified.
- To study isotonic and isometric contractions and various movements of the skeleton.

- skeletal musculature models
- atlas and dissection guide in manual
- dissecting instruments
- dissecting trays
- protective eyewear
- fetal pigs
- weights
- rubber or latex gloves
- microscopes
- slides of skeletal muscle

⚠ *SAFETY ALERT!*

Preserved specimens contain chemicals that are potentially irritating to the skin and eyes. Wear gloves and protective eyewear when dissecting. Dissecting instruments are sharp. Take care not to puncture or cut your skin. Carry the microscopes with two hands. Slides are made of glass. Even when not broken, they are sharp. In the event that any glass is broken, inform the instructor to seek guidance concerning clean up procedures.

▌▌▌▌ Introduction

As was briefly discussed in Exercise 4, muscle tissue is unique because of its ability to **generate a force**. For the most part, **skeletal muscles** generate these forces under **voluntary** control. They most often work with the skeleton to bring about movement. This includes walking, running, and any movements of the torso, hands, and arms. Skeletal muscle is also responsible for other diverse activities such as breathing, smiling, speaking, and maintaining posture and balance. In addition, skeletal muscle can generate heat through rapid contractions commonly known as "shivering." Muscle tissue is also a storehouse for **glycogen**, a molecule used for energy production. In this lab, we focus on many of the superficial (on the surface) and a few of the deeper skeletal muscles that bring about such actions.

Naming and Identifying Muscles

The term **muscle** usually refers to a whole muscle such as the familiar biceps and triceps (**Figure 12-1**). Muscles are attached to other muscles and bones by way of tendons. In order for a muscle to bring

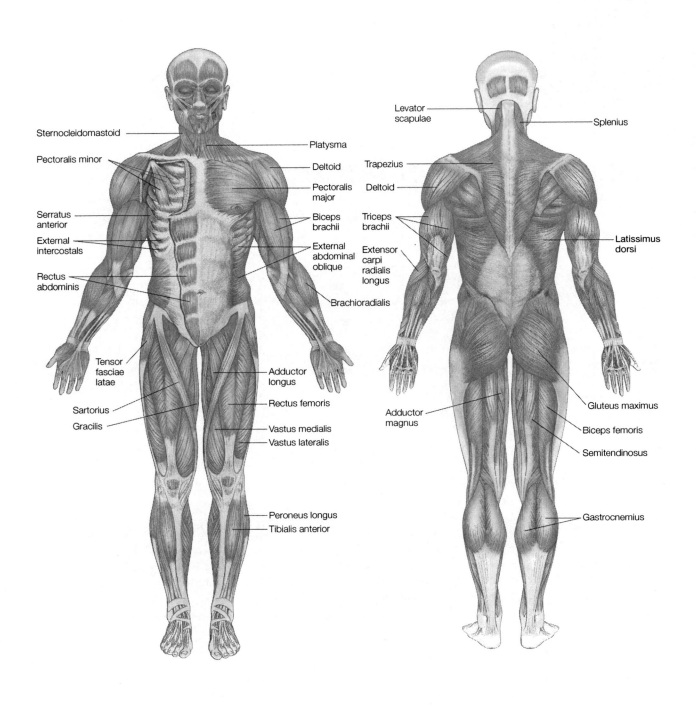

Sternocleidomastoid

Pectoralis minor

Serratus anterior

External intercostals

Rectus abdominis

Tensor fasciae latae

Sartorius

Gracilis

Platysma

Deltoid

Pectoralis major

Biceps brachii

External abdominal oblique

Brachioradialis

Adductor longus

Rectus femoris

Vastus medialis

Vastus lateralis

Peroneus longus

Tibialis anterior

Levator scapulae

Splenius

Trapezius

Deltoid

Triceps brachii

Extensor carpi radialis longus

Latissimus dorsi

Adductor magnus

Gluteus maximus

Biceps femoris

Semitendinosus

Gastrocnemius

Figure 12-1 Front and back view of skeletal muscles.

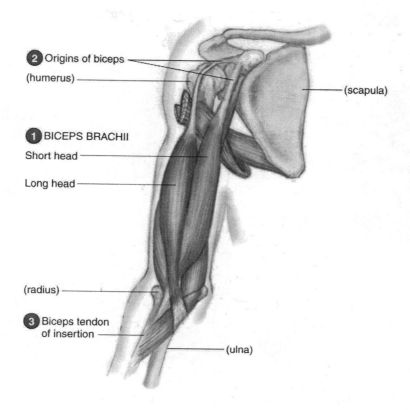

FRONT OF THE ARM
(with palm facing forward)

2 Origins of biceps

(humerus)

(scapula)

1 BICEPS BRACHII

Short head

Long head

(radius)

3 Biceps tendon
of insertion

(ulna)

Figure 12-2 The biceps brachii muscle showing origin and insertion.

about an action, it usually has to cross over a joint. The tendon that attaches to the bone that does not move during the action is called the **origin** (**Figure 12-2**). Conversely, the tendon attached to the movable bone is called the **insertion**.

Muscles can be named according to their location, shape, action, origin, insertion, and size. Many names will be derived from the ancient languages of Latin and Greek and therefore may seem a bit odd. For instance, **gluteus maximus** is Latin for the large (maximus) muscle in the buttocks (gluteal) region. The **masseter** is a large muscle in the lower jaw. The name is from the Greek word *maseter*, which means to chew.

Muscles are layered. That is, many are superficial while others lie deep. You will see this during dissection.

Muscle Fibers

Muscle cells are also called **muscle fibers** because they are long cylinders unlike the generalized, spherical animal cell you studied in Exercise 3. They are comprised of even smaller fibers, **myofibrils**, built from two proteins, **actin** and **myosin** (**Figure 12-3**). These proteins are organized into small units called **sarcomeres**. There are thousands of sarcomeres in every muscle cell. The actin and myosin slide past each other during muscle contraction. This shortens, or "contracts," the sarcomere and generates a force.

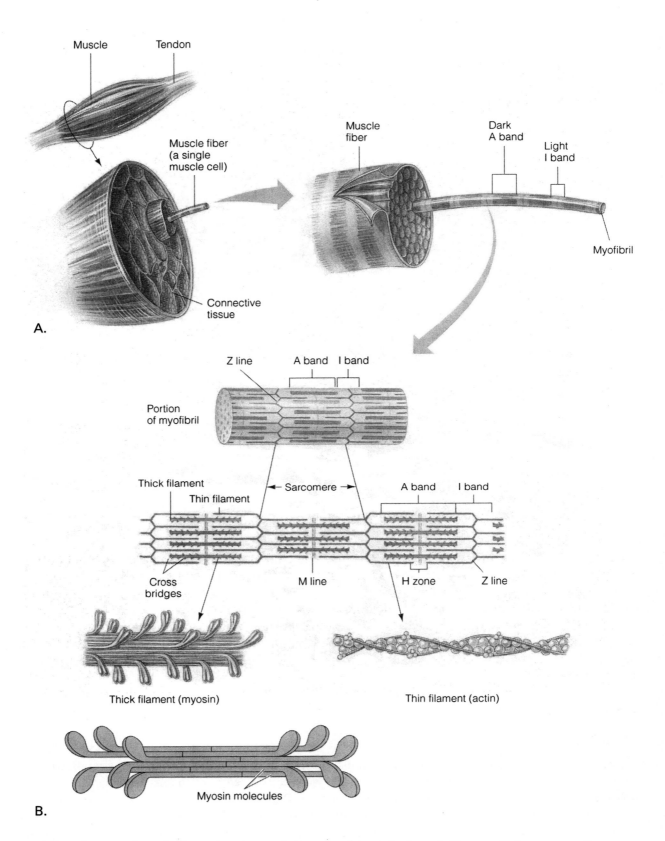

Muscle

Tendon

Muscle fiber
(a single
muscle cell)

Connective
tissue

A.

Muscle
fiber

Dark
A band

Light
I band

Myofibril

Z line

A band   I band

Portion
of myofibril

Thick filament

Thin filament

← Sarcomere →

A band   I band

Cross
bridges

M line

H zone

Z line

Thick filament (myosin)

Thin filament (actin)

Myosin molecules

B.

Figure 12-3 Structure of muscle fiber and sarcomere. A. Structure of the skeletal muscle fiber, myofibril, sarcomere. B. Structure of myosin filaments.

procedure

1 Models and Diagrams

Using the models, Figure 12-1, and your textbook, identify the following muscles:

masseter	levator scapulae	splenius
pectoralis major	pectoralis minor	deltoid
rectus abdominus	biceps brachii	triceps brachii
trapezius	sternocleidomastoid	external oblique
latissimus dorsi	gluteus maximus	biceps femoris
sartorius	gracilis	adductor longus
vastus medialis	vastus lateralis	rectus femoris
gastrocnemius	semitendinosus	peroneus longus

TABLE 12-1 lists the origin, insertion, and action of selected muscles.

TABLE 12.1	Origin, Action, & Insertion of Muscles		
Muscle	**Origin**	**Insertion**	**Action**
Masseter	Zygomatic Bone	Mandible	Elevates Mandible
Biceps Brachii	Scapula	Radius	Flexes Elbow
Deltoid	Clavicle & Scapula	Humerus	Abduct Shoulder
Triceps Brachii	Scapula & Humerus	Ulna	Extends Elbow
Gracilis	Pubis	Tibia	Adduct & Flex Hip
Sartorius	Ilium	Tibia	Flexes Knee & Hip
Vastus Medialis	Femur	Tibia	Extends Knee
Rectus Femoris	Ilium	Tibia	Flexes Hip
Semitendinosus	Ischium	Tibia	Flexes Knee
Gastrocnemius	Femur	Calcaneus	Flex Ankle & Knee

2 Dissection

Using fetal pigs and Appendix 2, identify as many of the following muscles as possible:

masseter	rectus abdominus	external oblique
deltoid	biceps brachii	triceps brachii
latissimus dorsi	brachialis	sartorius
gracilis	gastrocnemius	semitendinosus
gluteus medius	biceps femoris	tensor fascia latae

3 Muscle Contraction and Strength

Two basic types of muscle contraction exist: **isotonic** and **isometric**. Isotonic ("iso" = same; "tonic" = tension) contractions are those that usually bring about movement. This can include lifting weights and walking. During these actions, the tension on the muscle remains the same but the fibers actually shorten.

During isometric contraction, the length of the fibers remains constant, but the tension increases. This is what you experience when you hold an object at your side for an extended period. This type of contraction also accounts for muscle tone and posture maintenance.

Some actions only require a small amount of force to be generated. Therefore, not all of the fibers in a given muscle will always contract. In order to increase the force, the muscle will recruit more fibers.

3 Movements: Flexion-Extension-Adduction-Abduction

You have surely heard the phrase "flex your muscles." But what does it mean? The term *flexion* refers to an action that decreases the angle of a joint. This occurs during the bicep (biceps brachii) curl in weight training. The opposing action would be the **extension** of the arm, which increases the angle at the joint. This is accomplished with the triceps brachii. The biceps and triceps are said to **antagonistic**: they bring about opposite actions (**Figure 12-4**).

Adduction is a movement that brings the arms or legs closer to the body, whereas **abduction** will take them away from the body (**Figure 12-5**).

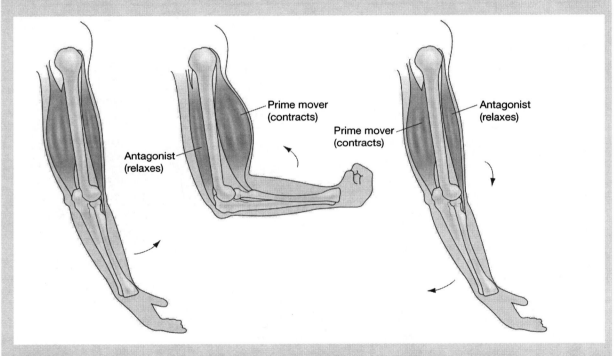

Prime mover (contracts)

Prime mover (contracts)

Antagonist (relaxes)

Antagonist (relaxes)

Antagonist (relaxes)

Figure 12-4 Muscle actions. Biceps flex, and triceps extend. They are antagonistic.

3 Slide of Skeletal Muscle

a. Observe the slide under low power.

b. Adjust the lighting and increase the magnification to see the striations.

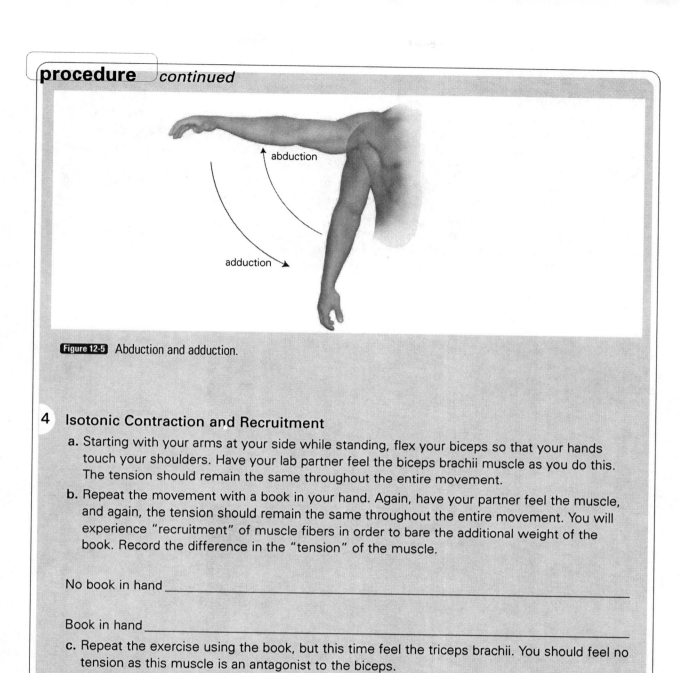

Figure 12-5 Abduction and adduction.

4 Isotonic Contraction and Recruitment

a. Starting with your arms at your side while standing, flex your biceps so that your hands touch your shoulders. Have your lab partner feel the biceps brachii muscle as you do this. The tension should remain the same throughout the entire movement.

b. Repeat the movement with a book in your hand. Again, have your partner feel the muscle, and again, the tension should remain the same throughout the entire movement. You will experience "recruitment" of muscle fibers in order to bare the additional weight of the book. Record the difference in the "tension" of the muscle.

No book in hand _____

Book in hand _____

c. Repeat the exercise using the book, but this time feel the triceps brachii. You should feel no tension as this muscle is an antagonist to the biceps.

4 Isometric Contraction

a. With your arms at your side, hold a book in one of your hands for 3 to 5 minutes.

b. At 1-minute intervals, have you partner feel the tension in both the triceps and biceps. Record your findings.

Tension after 1 minute _____

Tension after 3 minutes _____

Tension after 5 minutes _____

Muscular Dystrophy

Muscular dystrophy is an inherited disease that is characterized by a progressive degeneration of muscle cells. Those afflicted produce very little or no dystrophin. This protein is responsible for helping to maintain the structural integrity of muscle cell membranes. Without dystrophin the cell membranes are easily are ruptured. This leads to death of the muscle cells.

Symptoms include falling while walking or running. This usually occurs in very early childhood, around the age of 3 years. By 10 to 12 years old, most cannot walk. Death is imminent by the age of 30 because of cardiac and respiratory failure.

Because the gene that codes for dystrophin is carried on the X chromosome, the disease afflicts mostly males. We discuss inheritance in Exercise 16.

Name: _____   Lab Section: _____

|||||| Review Questions

1. How can skeletal muscle rapidly generate heat?

   _____

   _____

2. Which two proteins play a major role in muscle contraction?

   _____

   _____

3. What molecule is stored in muscle tissue that can be used to produce energy?

   _____

   _____

4. Two muscles that bring about opposite effects are known as what?

   _____

   _____

5. Define and distinguish between the "origin" and "insertion."

   _____

   _____

6. Why is skeletal muscle also referred to as voluntary muscle?

   _____

   _____

7. Define flexion.

   _____

   _____

Name: _____   Lab Section: _____

8. Label the diagram of the musculature system.

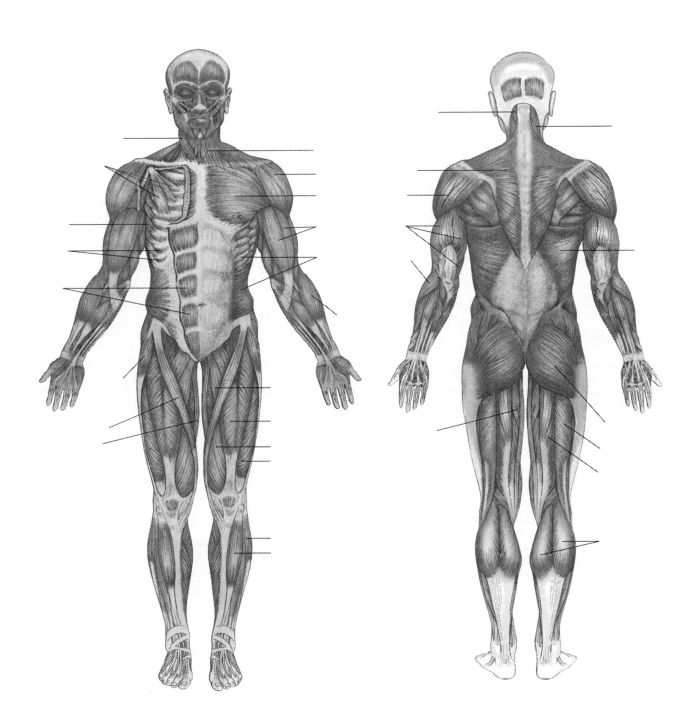

Nervous System 1: Brain and Spinal Cord

objectives

- To identify the structures of the brain and spinal cord and their functions
- To identify the structures of a neuron under the microscope
- To identify cranial nerves and their functions

materials

- sheep brains
- human brain models
- dissecting trays
- dissecting instruments
- spinal cord models
- slides of neurons
- microscopes
- rubber or latex gloves
- protective eyewear

⚠ SAFETY ALERT!

Preserved specimens contain chemicals that are potentially irritating to the skin and eyes. Do not handle specimens without gloves or protective eyewear. Dissection instruments are sharp. Take care not to puncture or cut your skin. Handle the microscopes carefully, and always carry them with two hands. Remember that the slides are made of glass. Even when not broken, the edges are sharp. In the event that a slide is broken, inform the instructor to seek guidance concerning cleanup procedures.

⦀⦀ Introduction

As discussed in Exercise 5, the **neuron**, or **nerve cell**, is the fundamental unit in the nervous system (**Figure 13-1**). Neurons are responsible for conducting impulses in all of the structures in the nervous system. Conduction is one way. Neurons receive information at the **dendrites** and send it along the **axon**.

The nervous system consists of the **brain**, **spinal cord**, and all **nerves** in the body. The brain and spinal cord together form the **central nervous system** (**Figure 13-2**). It is so named because all information is processed through (into and out of) the spinal cord, brain, or both.

Nerves make up the **peripheral nervous system**. Nerves are actually bundles of neurons, or nerve fibers. **Cranial nerves** communicate directly with the brain, whereas **spinal nerves** arise from the spinal cord to communicate with the entire body.

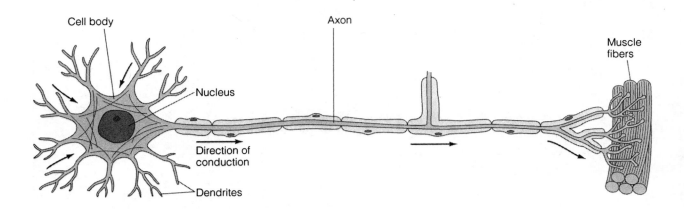

Figure 13-1 Structure of a neuron.

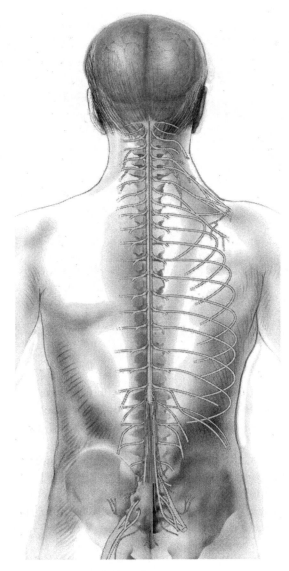

Figure 13-2 The central nervous system, brain, and spinal cord.

procedure

1 The Brain

Using **Figure 13-3** and **Figure 13-4**, sheep brain specimens, and the human brain models, identify the structures of the brain listed.

Cerebrum: The seat of higher intellectual capacity such as learning, memory, and speech; has several lobes: frontal, parietal, temporal, and occipital.

Cerebellum: Coordinates skeletal muscle activity.

Olfactory bulb and tract: Carries neurons for sense of smell.

Optic nerve, optic chiasma and optic tract: Carries information associated with vision.

Pons: Helps to regulate breathing and is involved in the control of facial muscles.

Medulla: Regulation of breathing and heart rate and relay for reflexes.

Thalamus: Relay center for all sensory input except olfaction (smell).

Corpus callosum: Connects the two cerebral hemispheres.

Pineal gland: Produces the hormone melatonin that controls circadian rythyms ("biological clock").

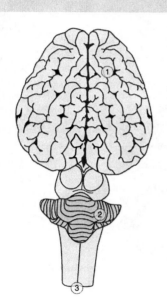

A.

1. cerebrum
2. cerebellar hemisphere
3. spinal cord

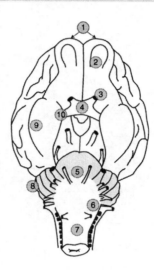

B.

1. frontal lobes
2. olfactory bulb
3. optic nerve
4. optic chiasma
5. pons
6. medulla oblongata
7. spinal cord
8. cerebellum
9. temporal lobe
10. optic tract

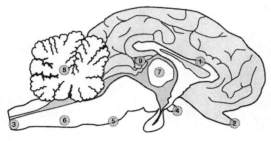

C.

1. corpus callosum
2. olfactory bulb
3. spinal cord
4. optic chiasma
5. pons
6. medulla oblongata
7. intermediate mass of thalmus
8. cerebellum
9. pineal gland

Figure 13-3 Sheep brain. (A) Top view. (B) Bottom view. (C) Medial view of half brain.

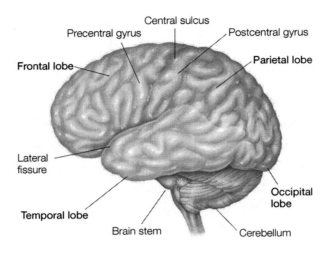

Figure 13-4A Whole human brain.

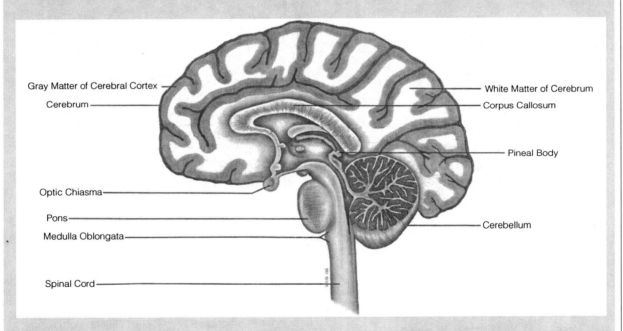

Figure 13-4B Half human brain.

2 Cranial Nerves

Twelve pairs of cranial nerves arise from the brain and innervate the region of the head and neck. They are named as well as numbered (**TABLE 13-1**). Cranial nerves can carry sensory or motor neurons or both. Cranial nerve X, the vagus nerve, is exceptional because it reaches into the abdomen.

Using brain models, **Figure 13-5**, and Table 13-1, identify the location and functions of the cranial nerves.

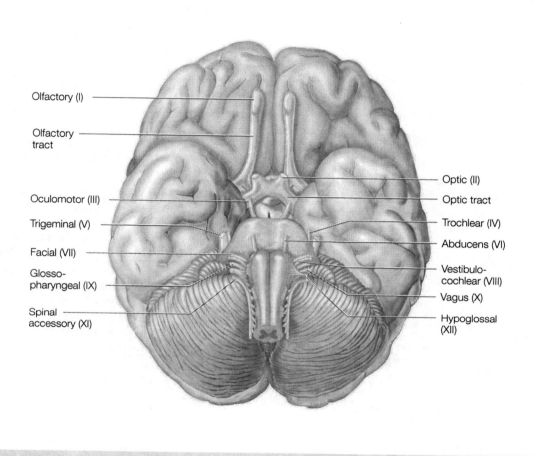

Olfactory (I)
Olfactory tract
Oculomotor (III)
Trigeminal (V)
Facial (VII)
Glosso-pharyngeal (IX)
Spinal accessory (XI)
Optic (II)
Optic tract
Trochlear (IV)
Abducens (VI)
Vestibulo-cochlear (VIII)
Vagus (X)
Hypoglossal (XII)

Figure 13-5 Bottom of human brain detailing cranial nerves.

3 Spinal Cord

The spinal cord communicates with the entire body by way of spinal nerves. They arise along the entire spinal cord and consist of **dorsal** (back) and **ventral** (front) roots (**Figure 13-6**). **Sensory** neurons, bringing signals into the spinal cord, are carried in the dorsal root. **Motor** neurons leaving the spinal cord are carried in the ventral root.

Using Figure 13-6 and spinal cord models, identify the following structures:

Gray matter: Mostly cell bodies and dendrites.

White matter: Mostly axons.

Central canal: Contains cerebrospinal fluid.

Dorsal root: Carries sensory neurons.

Ventral root: Carries motor neurons.

TABLE 13-1	**Cranial Nerves: Names, Numbers, and Functions**

Name and Number	Origin	Function
I. Olfactory	Cerebral cortex	Sense of smell
II. Optic	Thalamus	Sense of sight
III. Oculomotor	Midbrain	Eye movement, focusing, and pupil diameter
IV. Trochlear	Midbrain	Eye movement
V. Trigeminal	Pons	Sensation from head, face, and mouth and chewing
VI. Abducens	Pons	Eye movement
VII. Facial	Pons	Facial sensation and expression, taste (anterior two thirds of tongue), tear secretion and salivation
VIII. Vestibulocochlear	Pons and Medulla oblongata	Hearing, balance, and equilibrium
IX. Glossopharyngeal	Medulla oblongata	Swallowing, taste (posterior third of tongue), salivation, sensation from soft palate and pharynx, blood pressure, and dissolved blood gas concentration
X. Vagus	Medulla oblongata	Activity and sensation for parts of ear, diaphragm, pharynx, respiratory tract, gastrointestinal tract (esophagus through colon), and heart
XI. Accessory	Medulla oblongata	Head movement, voluntary component of swallowing, control of vocal cords
XII. Hypoglossal	Medulla oblongata	Voluntary tongue movements, speech, and swallowing

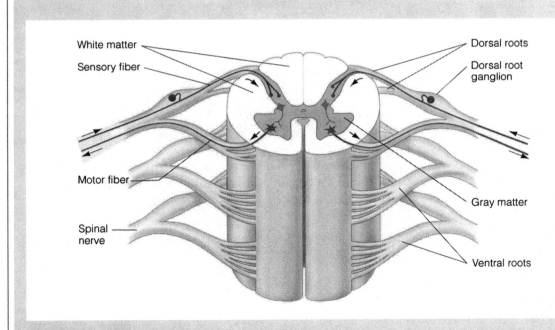

Figure 13-6 Cross-section of spinal cord showing dorsal and ventral roots.

4 Neuron Slide

a. Scan using low power under the microscope to view the entire neuron.

b. View under higher magnification, and identify the cell body, dendrites, and axon.

CLINICAL CONSIDERATIONS

Parkinson's Disease

Parkinson's disease is characterized by a progressive loss of control over skeletal muscle movements. This loss is caused by the degeneration of neurons deep in the brain that play a major role in initiating skeletal muscle movements.

The most common symptom exhibited is an uncontrollable shaking called a **tremor**. This is the result of involuntary muscle contraction interfering with voluntary movements.

The disease is also characterized by **bradykinesia** and **hypokinesia**. Bradykinesia is the extreme slowness in movement. Again, this is caused by the failure to properly initiate movements. Simple tasks such as opening a door or reaching for a glass can take several minutes. Hypokinesia is the condition in which the range of motion is reduced. For instance, one of the diagnostic tools used to identify Parkinson's is to study a patient's handwriting to see whether letters and words have progressively gotten smaller or even illegible.

Those affected with Parkinson's are usually diagnosed at about the age of 55 to 65 years. The cause is yet unknown, but some environmental factors are suspected. These include toxic chemicals such as carbon monoxide and organic chemicals such as acetone and pesticides; however, research has been focused on the possibility that genetic makeup may play a major role in susceptibility.

Name: _____   Lab Section: _____

|||||| Review Questions

1. Which root of a spinal nerve carries motor neurons?

   _____

2. Which structure in the brain connects the two cerebral hemispheres?

   _____

3. Which part of an axon receives information?

   _____

4. The two optic nerves cross at what junction?

   _____

5. What is the transition between the brain and spinal cord?

   _____

6. Which sense is served by cranial nerve I?

   _____

7. White matter is comprised mostly of what?

   _____

8. Which region of the brain coordinates skeletal muscle movements?

   _____

9. Draw and label the structure of a neuron.

Name: _____ Lab Section: _____

10. Label the diagrams of the brain and spinal cord.

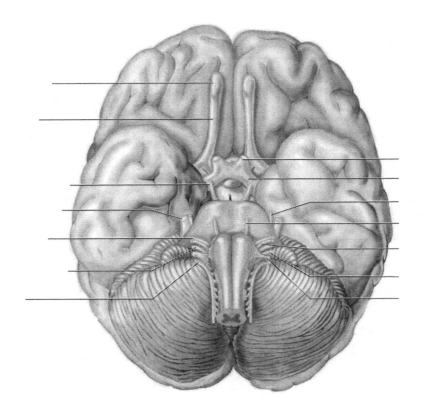

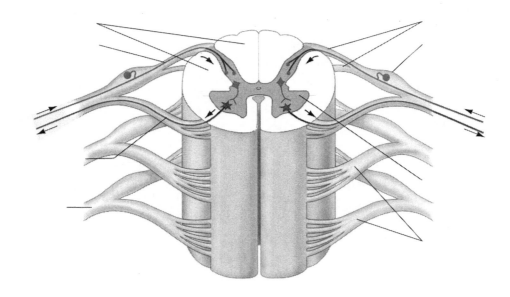

Nervous System 2: Senses

objectives

- To identify the structures associated with sensory input: eye, ear, tongue, and skin
- To test general and special sensations

materials

- cow eyes
- human eye models
- human ear models
- dissection trays
- dissection instruments
- rubber or latex gloves
- protective eyewear
- rubber mallets
- dissection pins
- scissors
- tuning forks
- 10% sugar solution
- 10% salt solution
- vinegar or lemon juice
- 5% quinine solution
- onions, apples, celery, and potatoes
- sterile cotton swabs
- small knives

▌*SAFETY ALERT!*

Preserved specimens contain chemicals that are potentially irritating to the skin and eyes. Do not handle specimens without gloves or protective eyewear. Dissection instruments are sharp. Take care not to puncture or cut your skin. Proceed with caution and respect when using human subjects.

⦚⦚⦚ Introduction

Sensory organs include the **eyes, ears, tongue,** and **nose**. They are associated with the "special" senses **vision, hearing, taste,** and **smell,** respectively. The "general" senses are associated with the **skin** and include **touch, pressure, temperature,** and **pain**.

External stimuli is received by these organs, converted into nervous impulses, and directed to the central nervous system.

In this exercise, you will examine the structure of the sense organs and various aspects of the senses they control.

procedure

1 The Eye and Vision

The eye is a complex structure responsible for capturing light rays and forming images. The images are sent as impulses along the optic nerve and into the brain. This accounts for the sense of vision.

Using **Figure 14-1** and **Figure 14-2** and human eye models, identify the structures and associated functions of the eye listed here:

Cornea: connective tissue covering; functions in protection and rough focusing of light rays

Pupil: opening to the lens

Iris: pigmented (colored) portion of the eye; muscle that controls the size of the pupil

Lens: focuses light rays onto the retina

Retina: region where light sensitive photoreceptors are found and images are formed

Ciliary body: smooth muscle that controls the shape of the lens

Cones: photoreceptors responsible for color vision in bright light; concentrated mostly in the center of the retina in a region called the fovea

Rods: Photoreceptors responsible for vision in dim light; very sensitive in low levels of light; concentrated mostly on the periphery (edge) of the retina

Sclera: connective tissue that serves to protect and provide site of attachment for eye muscles.

Optic nerve: transmits impulses from the retina to the brain

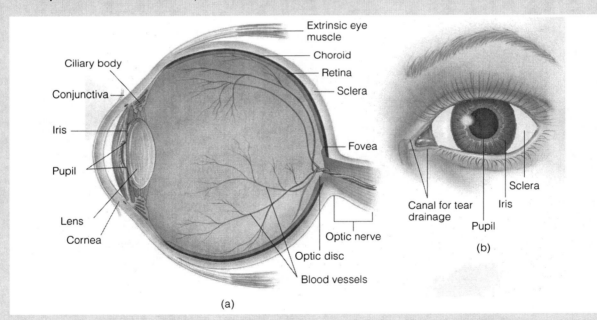

Figure 14-1 Anatomy of the Eye.

The Blind Spot

The region where the optic nerve exits the retina contains no photoreceptors. Therefore, if light rays fall on this region, no image is formed, and a **blind spot** is perceived. You have no blind spot in normal vision because the left and right eyes compliment each other. However, if you use only one eye, you can demonstrate the existence of the blind spot.

Figure 14-2 Cross and dot.

a. Hold your lab manual with Figure 14-2 about 40 cm, or 16 inches, in front of your eyes.

b. Close your right eye, and with your left eye, focus on the circle.

c. Slowly move the manual toward you until the plus sign disappears.

d. Have your partner record the distance between the manual and your eye at the point where it disappears.

_____

e. Repeat the procedure closing the left eye and focusing on the plus sign with the right eye. Have your lab partner record the distance between the manual and your eye when the circle disappears.

_____

Accommodation

The shape of the lens of the eye changes in response to viewing objects at different distances. This is known as **accommodation**. When viewing near objects, the ciliary muscles contract, causing the lens to become somewhat short and rounded. Conversely, when the muscles are relaxed to view distant objects the lens becomes flattened and elongated. As we age, we slowly lose the contractility of these muscles and therefore the ability to "accommodate" for near objects.

a. Hold a pencil with the point facing up at arms length.

b. Close one eye. With the other eye, focus on the point as you slowly bring the pencil toward you until the point is no longer in focus.

c. When the point is no longer in focus, have your lab partner measure the distance in centimeters between your eye and the pencil. Repeat for the other eye. Record your results.

Right Eye _____

Left Eye _____

2 The Ear, Hearing, and Equilibrium

The ear is actually divided into three regions: outer ear, middle ear, and inner ear (**Figure 14-3**). Structures of the ear are responsible for converting external sound waves into organized "vibrations" that can then be converted into nervous impulses. The inner ear also helps maintain equilibrium and balance.

Using Figure 14-3 and the human ear models, identify the structures and functions of each

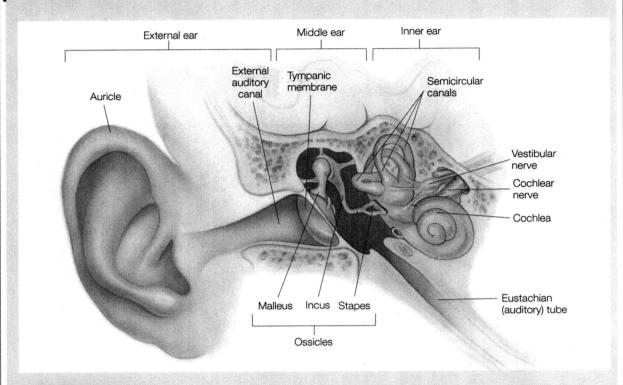

Figure 14-3 Structures of the ear.

region listed here.

Outer Ear

   Auricle: captures sound waves

   External auditory canal: directs sound waves to the eardrum

Middle Ear

   Tympanic membrane (eardrum): vibrates in response to sound waves

   Ossicles (ear bones): interprets and transmits vibrations to the cochlea

Inner Ear

   Cochlea: converts waves into nerve impulses

   Cochlear nerve: carries impulses into the brain

   Semicircular canals: help maintain balance and equilibrium

Hearing Tests
Simple tests can reveal hearing loss, or deafness. **Nerve** deafness results from physical damage to parts of the inner ear or cochlear nerve. **Conduction** deafness results from the obstruction of the flow of sound waves. Affected regions can include the eardrum and the ossicles.

Weber Test for Nerve Deafness
   a. Strike a tuning fork on a table and press the end of the handle to middle of your forehead.

   b. If the tone is heard in the middle of the head, you have equal hearing or hearing loss in both ears. If nerve deafness occurs in one ear, the tone will be heard in the other ear. If there is any conduction deafness, the tone will still be heard in this test because the tissues of the

procedure *continued*

skull will transmit the vibrations to the cochlea. Record your results.

Rinne Test for Conduction Deafness

a. Strike a tuning fork on a table and press the end of the handle on your mastoid process until the sounds fades away.

b. After the sounds fades, strike the tuning fork again and place the tines in front of the auditory canal. If the sound was louder and lasted longer when pressed on the mastoid process, there is possibly a conduction loss. Record your results.

---

3 Taste and Smell

The senses of taste and smell are both involved in detecting chemicals in the environment. This is known as **chemoreception**. Molecules in the air and food will excite the neurons in the olfactory bulb in the nasal cavity and tastes buds on the tongue.

As you are probably aware from your own experience, taste and smell have evolved together as a combined sense. This is due to the transporting of molecules between the mouth and nose. This is why when your sinuses are clogged you cannot taste as well as you do normally.

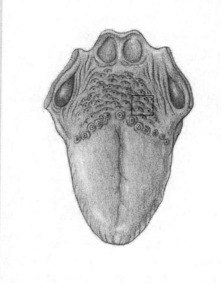

Figure 14-4 Tongue.

Determining Distribution of Taste Buds

Four major taste sensations exist: **sweet**, **sour**, **bitter**, and **salty**. Taste buds for all of these are found on the tongue but are not distributed evenly. In this exercise you will attempt to plot the distribution.

a. Dip a sterile cotton swab into the 10% sugar solution.

b. Touch the swab onto the front, back, middle, and sides of your partners tongue.

c. Using the diagram of the tongue (**Figure 14-4**), indicate where the sweet sensation was experienced.

d. Repeat the procedure using the salt solution, vinegar or lemon juice, and quinine. Have your partner rinse his or her mouth after each solution for best results.

Taste and Smell

In this exercise you will examine the intimate relationship between taste and smell.

a. Cut the apples, potatoes, onions, and celery into bite-sized pieces.

b. While closing his or her eyes and holding his or her nose, have your lab partner chew a piece of potato or apple. These foods have similar textures; therefore, they will not know which it is by touch alone.

c. Have your partner attempt to determine which it was and record the results.

d. Repeat using the onions and celery.

_____

4 Sensations of the Skin

Two-Point Discrimination

Different regions of the skin contain varying densities of "touch" receptors. This accounts for the differences in sensitivities and can be shown by performing a **Two-Point Discrimination** test. In this test you will measure your lab partners' ability to discriminate between two points of pressure applied to the skin.

Using a pair of scissors or dissecting pins, you will attempt to find the minimal distance required to detect that "two points" have been stimulated simultaneously. Do the test on the palm of the hand, fingertips, middle of the back, and forearm.

a. Start with two points about 5 mm apart.

b. If your partner can discriminate between these, shorten the distance. If he or she cannot, increase the distance.

c. Measure and record the minimal distance in millimeters.

Palm _____

Fingertips _____

Back _____

Forearm _____

Pressure and Adaptation

The skin contains neurons that function specifically to detect intensity of applied pressure. However, these receptors will stop responding to a stimulus after a while. This is known as **adaptation**.

a. Place a few coins on the inside of your lab partners' forearm.

b. Record how long it takes before your partner is no longer aware of the sensation of pressure (i.e., how long until they adapted).

_____

5 Reflexes

A **reflex** is an involuntary response to a stimulus. The patellar, or "knee jerk," reflex is a diagnostic tool used to test the function of spinal nerves. It is the most simple circuit in the nervous system. It consists of only two neurons: one sensory and one motor (**Figure 14-5**).

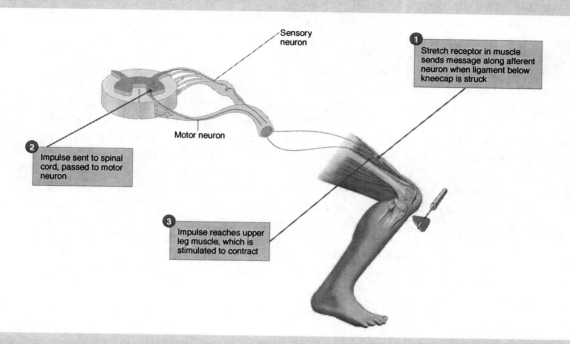

Sensory neuron

Motor neuron

1 Stretch receptor in muscle sends message along afferent neuron when ligament below kneecap is struck

2 Impulse sent to spinal cord, passed to motor neuron

3 Impulse reaches upper leg muscle, which is stimulated to contract

Figure 14-5 Knee jerk reflex.

Reflex Arc

a. Have your lab partner sit on a chair or lab bench with legs crossed.

b. Using a rubber mallet strike your partner in the region of the patellar tendon and record the results. Repeat for the other leg.

Right Leg _____

Left Leg _____

Nearsighted and Farsighted

Nearsightedness, or **myopia**, is characterized by the lack of ability to focus images from distant objects onto the retina. It is called nearsightedness because objects that are near to the eye can be focused. It can be caused by a slightly elongated eyeball or a lens that is too concave. These conditions result in the light rays from distant objects converging on a focal point in front of the retina. The light rays then scatter, resulting in a blurred image.

Farsightedness, or **hyperopia**, can result from an eyeball that is somewhat short and stout or from a lens that is too convex, or weak. As the name implies, distant objects are focused well but those nearer are not. In this condition the light rays from near objects attempt to focus beyond the retina. Therefore, the light rays that meet the retina are not yet focused, again resulting in a blurred image.

Options to restore normal vision include wearing traditional eyeglasses with corrective lenses or contact lenses. The latest technology employed to restore normal vision is laser surgery. In this procedure the cornea, which is the first structure to begin focusing light rays, is reshaped with a laser.

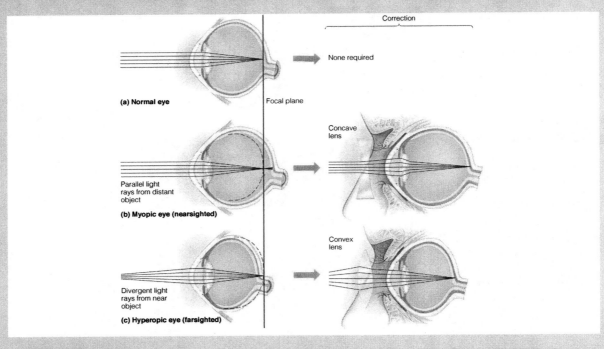

Figure 14-6A Vision problems.

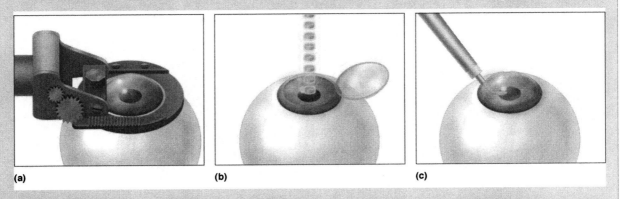

Figure 14-6B Vision corrective options: Lasik surgery. (a) Microkeratome slices off a thin layer of corneal tissue. (b) Laser burns away corneal tissue to correct eyesight. (c) Flap is restored, as is vision.

Name: _____ Lab Section: _____

IIIIII Review Questions

1. Which structure of the eye controls the size of the pupil?

   _____

   _____

2. What type of deafness would you experience if your eardrum was torn?

   _____

   _____

3. Which photoreceptors are responsible for "night" vision?

   _____

   _____

4. Name one sense involved in chemoreception?

   _____

   _____

5. Which of the three regions of the ear contains the ossicles?

   _____

   _____

6. Which structure in the eye is pigmented?

   _____

   _____

7. Neurons for the sense of smell are located where?

   _____

   _____

8. What is the connective tissue on the eye that serves as an attachment for eye muscles?

   _____

   _____

9. Define adaptation.

   _____

   _____

Name: _____ Lab Section: _____

10. Why do you not perceive a blind spot in normal vision?

_____

_____

11. Label the diagrams of the eye and ear.

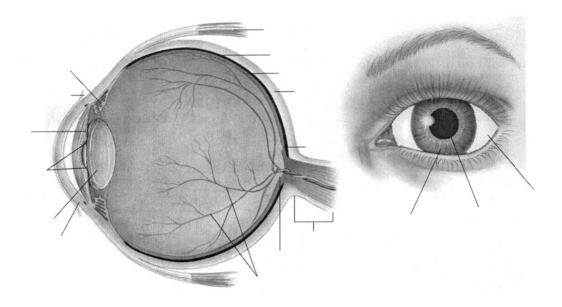

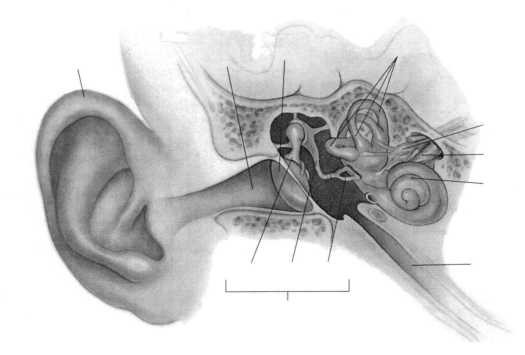

Urinary and Reproductive Systems

objectives

- To become familiar with the structure and function of the reproductive and urinary systems
- To study gamete formation through meiosis

materials

- models of human reproductive system
- models of human urinary system
- fetal pigs
- dissecting trays
- dissecting instruments
- protective eyewear
- latex or rubber gloves
- urinalysis kits
- pH paper

⚠ SAFETY ALERT!

Preserved specimens contain chemicals that are potentially irritating to the skin and eyes. Do not handle specimens without gloves or protective eyewear. Dissecting instruments are sharp. Take care not to puncture or cut your skin. Handle urine as if it were infectious—with great care .

⦚⦚⦚ Introduction

Historically, the **reproductive** and **urinary** systems have been taught together as the **urogenital** system. This is a logical approach because of the close anatomical association of the two systems.

The urinary system functions to maintain water balance by producing and eliminating the waste product **urine**. It includes the **kidneys, ureters, bladder,** and **urethra**.

The reproductive system functions to produce offspring. Male and female reproductive systems differ dramatically and are presented separately.

The process of **meiosis** will produce **gametes, sperm,** in males, and **eggs** in females.

procedure

1 Urinary System

Using **Figure 15-1**, models of human kidneys and the urinary system, appendix 2, and fetal pigs, identify the structures and their functions listed here:

Kidneys: There are two kidneys, one on either side of the back of the abdominal wall. The kidneys function to maintain water balance, filter blood, and produce urine. Each kidney empties urine into its own ureter.

Renal pelvis: collects urine and is continuous with the ureter.

Renal cortex: outer portion of the kidney involved in blood filtration.

Renal medulla: inner portion of the kidney; contains renal pyramids where urine is collected and renal columns that are rich in blood vessels for transport to and from the cortex.

Ureters: carry urine to the bladder.

Bladder: stores urine until it is eliminated.

Urethra: eliminates urine.

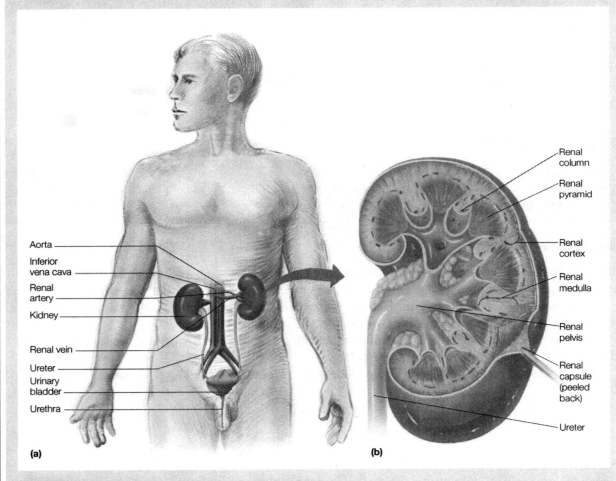

(a)

(b)

Renal column
Renal pyramid
Renal cortex
Renal medulla
Renal pelvis
Renal capsule (peeled back)
Ureter

Aorta
Inferior vena cava
Renal artery
Kidney
Renal vein
Ureter
Urinary bladder
Urethra

Figure 15-1 (A) Urinary system. (B) Cross-section of the kidney.

Urinalysis

Contents of urine can be studied and used in a medical diagnosis. Levels of chemicals such as glucose, albumin, and urea can indicate possible pathologies. The pH of the urine is also an important indicator.

a. Obtain fresh urine in a sterile container.

b. Using the pH strips, chemstix or strips, or any other material your instructor provides, perform the tests for glucose, albumin, urea, and pH.

c. Record your results.

pH: The normal range is 4.5 to 8.0._____

Glucose: This should not be present in high amounts._____

Protein (albumin): This should not be present._____

Urea: This should be present._____

2 Female Reproductive System

The reproductive systems of both females and males are responsible for producing gametes, or sex cells. The female reproductive system is more complex, however, because it also provides the site for fertilization (the union of egg and sperm) and growth of the fetus.

Using **Figure 15-2**, models of the female human reproductive system, fetal pigs, and appendix 2, identify all of the structures and associated functions listed here.

Ovaries: produce eggs (ova) and female sex hormones.

Uterine (fallopian) tubes: transport eggs to uterus.

Uterus: site of fetal development.

Vagina: receives penis and sperm during copulation; also serves as the birth canal.

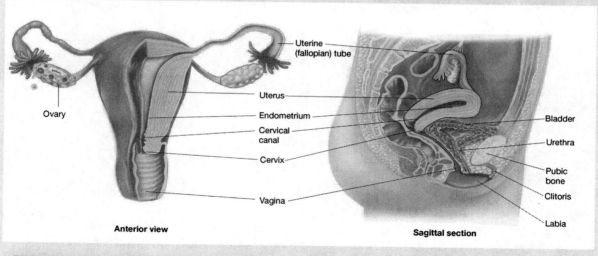

Anterior view   Sagittal section

Figure 15-2 Female reproductive system.

3 Male Reproductive System

The male reproductive system is built only to produce sperm and deposit it into the female system. Therefore, it is not as physiologically complex as the female system.

Using **Figure 15-3**, models of the male human reproductive system, fetal pigs, and appendix 2, identify all of the structures and associated functions listed here.

Testicles (testes): produce sperm and male sex hormones.

Epididymis: stores sperm.

Vas deferens: transports sperm to the urethra.

Prostate gland and bulbourethral gland: produces the fluid portion of the semen for nourishment of the sperm.

Urethra: transports sperm and urine to the outside of the body.

Penis: copulatory organ.

Scrotum: provides protection and temperature regulation for testicles.

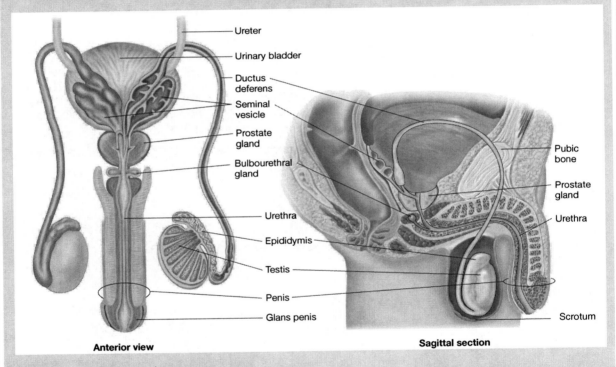

Anterior view   **Sagittal section**

Figure 15-3 Male reproductive system.

4 Gamete Formation and Meiosis

Sperm and eggs cells are known as **gametes**. They are formed through a process known as **gametogenesis, or meiosis**. Meiosis is a type of cell division very much like mitosis. However, mitosis involves one division that produces daughter cells that are **diploid**, or have the full chromosomal compliment found in the mother cell. Meiosis, on the other hand, exhibits two divisions and produces cells that are **haploid**, or have only half of the full number of chromosomes. This is important for sexually reproducing organisms so that when sperm and egg cells unite the full diploid number is restored.

All of the cells in your body have **46** (diploid) chromosomes, except for gametes, which have **23** (haploid). Fertilization restores the diploid number of 46.

Follow the stages of meiosis listed here using **Figure 15-4** and **Figure 15-5**. In humans, the diploid number is 46, and the haploid is 23. Four chromosomes represent the diploid number for simplification in the figures.

Meiosis I

Interphase I: chromosomes duplicate and now have four "double" chromosomes each with two "chromatids."

Prophase I: chromosomes pair up.

Metaphase I: chromosomes line up on equator of the cell.

Anaphase I: chromosomes split apart and leave two new cells each with "2" double chromosomes each with two chromatids. The chromosome number is now haploid, 2.

Telophase I: formation of two new "haploid" cells.

Meiosis II

Prophase II: nucleus disappears, again exposing chromosomes.

Metaphase II: chromosomes line up on equator of the cell.

Anaphase II: chromosomes split apart. This leaves two new haploid cells, each with 2 "single" chromosomes, or chromatids.

Telophase II: formation of four new haploid cells, or gametes.

5 **Human Development**

The union of the egg and sperm is known as fertilization, or the creation of a **zygote**. The zygote goes through a series of rapid mitotic divisions (**Figure 15-6**), with no increase in size, called **cleavage**. At the 16 cell stage it is called a **morula**. This stage is followed by the **blastocyst**. The blasocyst is what actually implants into the uterus. The embryo now begins to grow larger. The next stage is the **gastrula**. This stage exhibits the three primary germ layers of cells (**Figure 15-7**): **endoderm**, **mesoderm**, and **ectoderm**. Further along in embryonic development, these will give rise to all body tissues (**Figure 15-8**).

After implantation, the extraembryonic structures are formed. These allow for protection and exchange of gases, nutrients, and waste between the embryo and mother. Find these structures using **Figure 15-9** and **Figure 15-10** and any available models:

Allantois: Serves as the fetal urinary bladder and functions in gas exchange.

Chorion: Outermost membrane and carries umbilical vessels.

Amnion: Thin membrane surrounding the embryo and fetus. Filled with amniotic fluid that cushions and maintains temperature.

Umbilical Vessels: Gas exchange between fetus and mother.

Embryonic development lasts for about 2 months. The remaining 7 months is known as fetal development (**Figure 15-11** and **Figure 15-12**). **TABLE 15-1** lists major events during development.

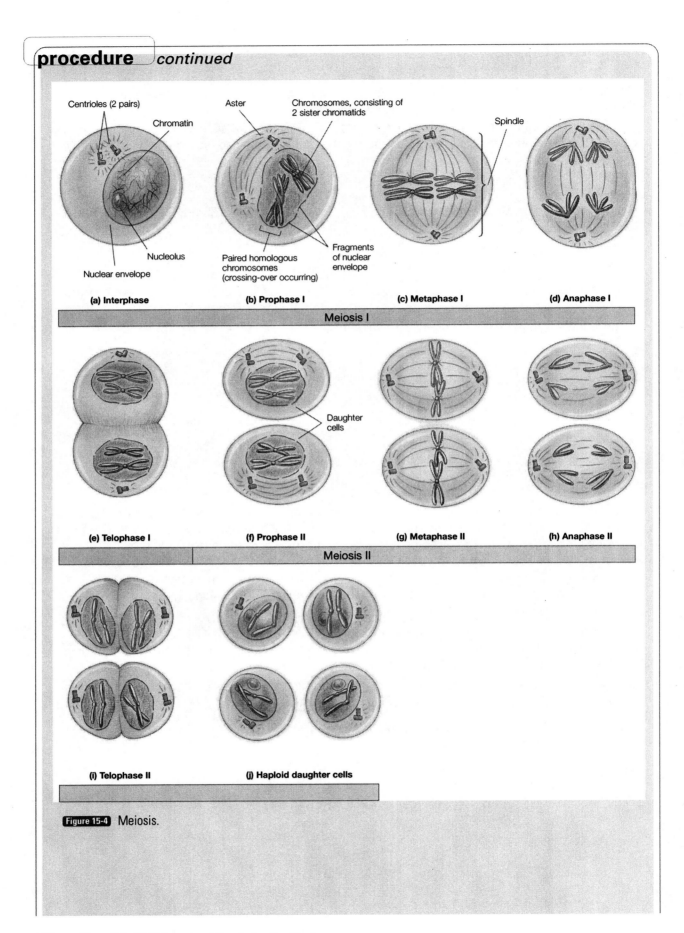

(a) Interphase **(b) Prophase I** **(c) Metaphase I** **(d) Anaphase I**

Meiosis I

(e) Telophase I **(f) Prophase II** **(g) Metaphase II** **(h) Anaphase II**

Meiosis II

(i) Telophase II **(j) Haploid daughter cells**

Figure 15-4 Meiosis.

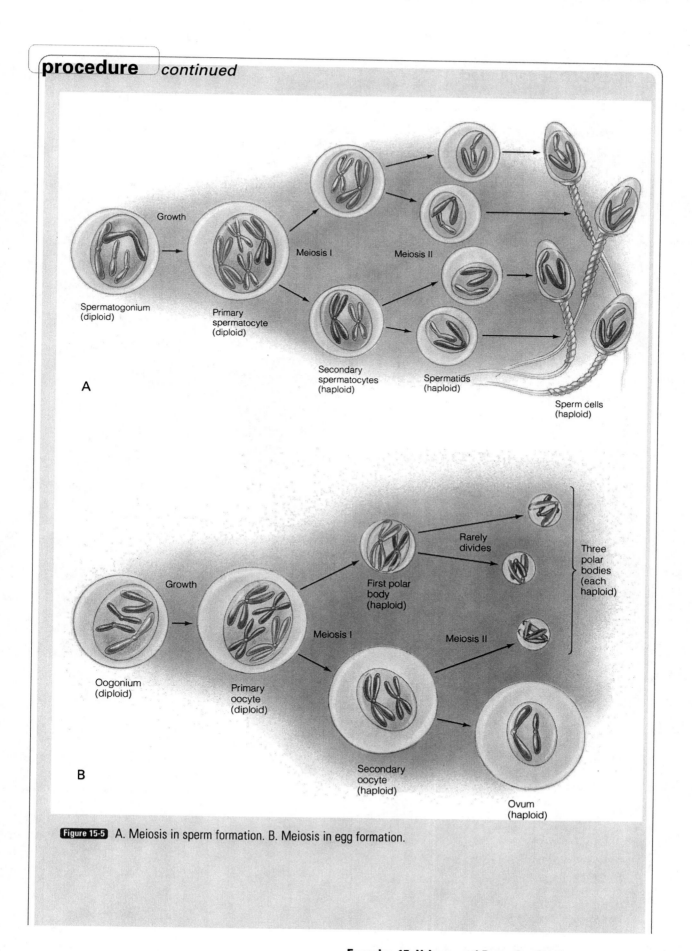

Figure 15-5 A. Meiosis in sperm formation. B. Meiosis in egg formation.

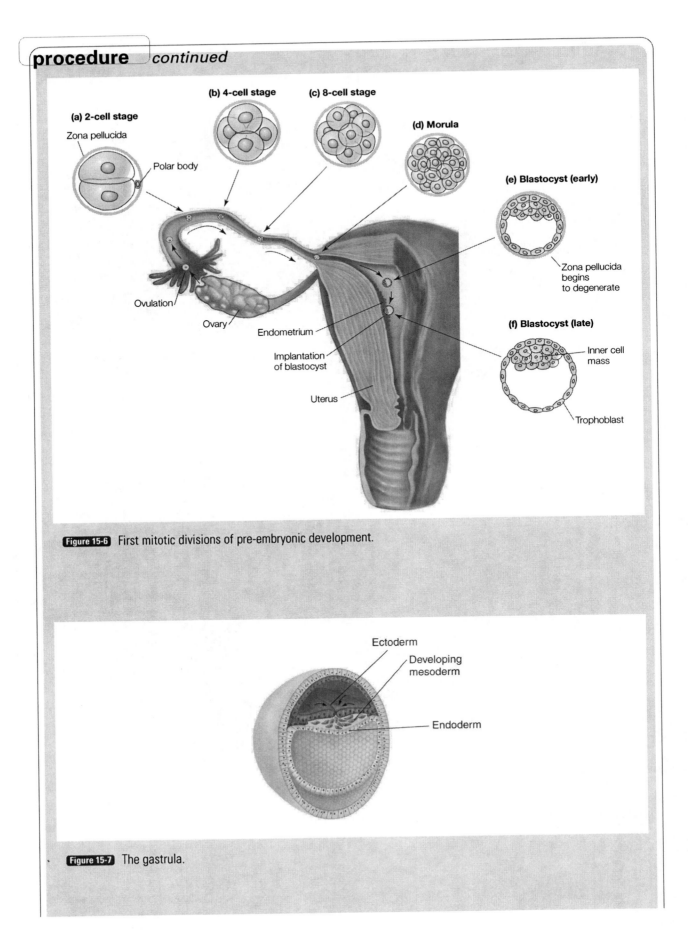

Figure 15-6 First mitotic divisions of pre-embryonic development.

Figure 15-7 The gastrula.

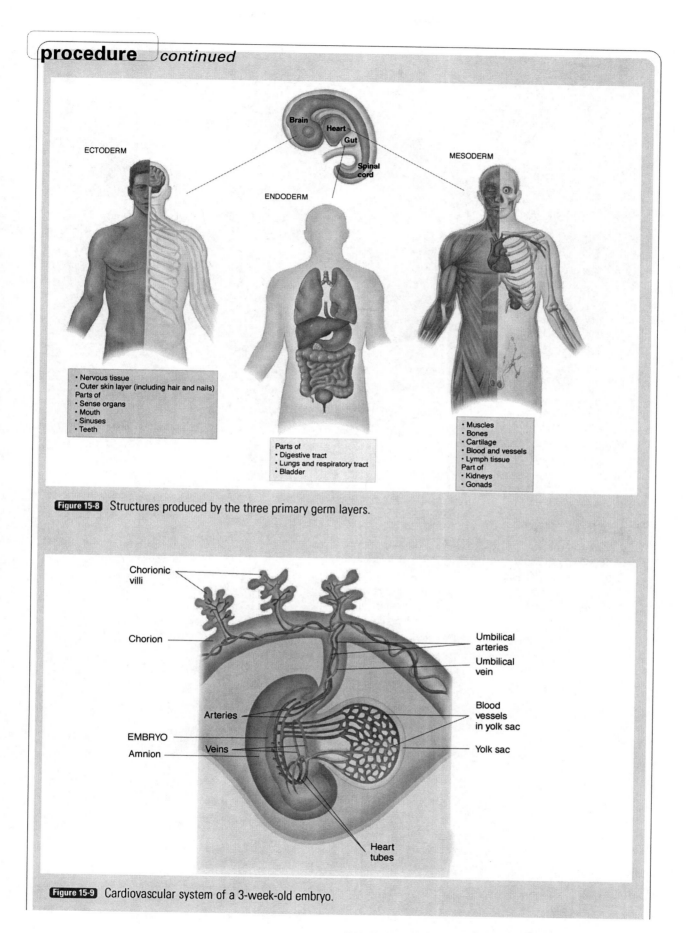

Figure 15-8 Structures produced by the three primary germ layers.

ECTODERM

ENDODERM

MESODERM

Brain
Heart
Gut
Spinal cord

• Nervous tissue
• Outer skin layer (including hair and nails)
Parts of
• Sense organs
• Mouth
• Sinuses
• Teeth

Parts of
• Digestive tract
• Lungs and respiratory tract
• Bladder

• Muscles
• Bones
• Cartilage
• Blood and vessels
• Lymph tissue
Part of
• Kidneys
• Gonads

Figure 15-9 Cardiovascular system of a 3-week-old embryo.

Chorionic villi

Chorion

Arteries

EMBRYO

Amnion

Veins

Umbilical arteries

Umbilical vein

Blood vessels in yolk sac

Yolk sac

Heart tubes

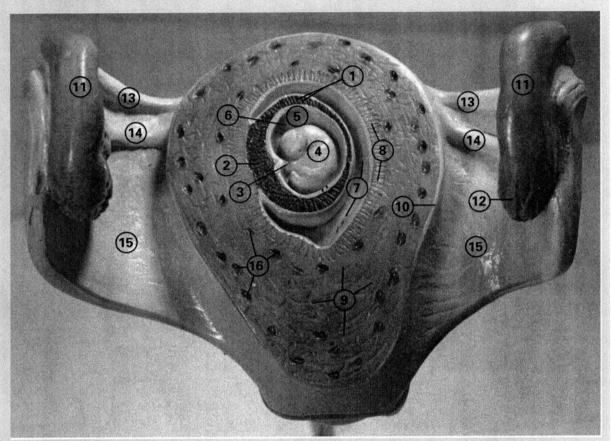

Figure 15-10 Embryonic and fetal development.

1. chorionic villi
2. developing placenta
3. developing umbilical cord
4. embryo
5. amnion
6. amniotic cavity
7. uterine cavity
8. endometrium of uterus
9. myometrium of uterus
10. perimetrium of uterus
11. uterine (fallopian) tube
12. fimbriated end of uterine tube
13. ovarian ligament
14. round ligament
15. broad ligament
16. blood vessels

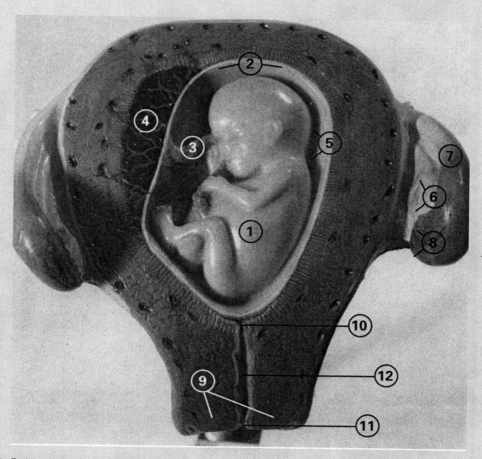

Figure 15-11 Embryonic and fetal development.

1. fetus
2. amnion
3. umbilical cord
4. placenta

5. amniotic cavity
6. ovary
7. uterine tube
8. fimbriated end of uterine tube

9. cervix
10. internal os
11. external os
12. cervical canal

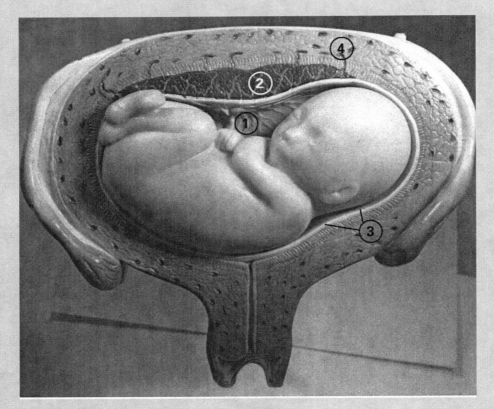

Figure 15-12 Fetal development, eight months, transverse position.

1. umbilical cord
2. placenta
3. amnion
4. myomentrium, stretched

procedure *continued*

TABLE 15-1	Embryological Development
Embryological Development	
Fertilization to the end of Week 8	**End of week 2:** Basic tissue types are present. **End of week 4:** Heart and nervous system are present and heart is beating. Limbs begin to develop. **End of week 6:** Fingers and toes are present. **End of week 8:** All systems are in place and bone is replacing the cartilaginous skeleton.
Fetal Development: 2 Months–Birth	
Week 8 to Birth	**End of month 4:** Skeleton and facial features prominent. Some joints formed. **End of month 6:** Body has grown to better match the large head. All systems rapidly growing. **End of month 8:** Fat is deposited below the skin. Testicles descend from abdomen into the scrotum in males. All systems almost completely developed. **End of month 9:** Body has grown in size and weight and is ready for birth.

CLINICAL CONSIDERATIONS

Infertility

Approximately 10% of all couples of reproductive age experience problems associated with conceiving, or **infertility**. Female infertility may be caused by a variety of factors. Ovarian problems may lead to the inability to produce fertile eggs. Obstruction of the uterine tubes will prevent sperm from fertilizing an egg. In addition, after fertilization the uterus may not be physiologically prepared to receive and nourish an embryo. This may result in a spontaneous, or early, miscarriage. A miscarriage at such an early stage is usually considered infertility. Poor nutrition and overall health may cause temporary reversible infertility. Women require a minimal amount of body fat to maintain regular menstrual cycles that include egg production followed by ovulation. Dieting and rigorous exercise may reduce body fat to levels too low to sustain menstrual cycles. Many female marathon runners experience such problems.

Male infertility, or **sterility**, is a result of the failure to fertilize a mature egg. The usual cause is a low sperm count or abnormal sperm morphology. Sperm is produced in the seminiferous tubules. The tubules are quite sensitive to infections, exposure to x-rays, and changes in temperature. Overexposure to heat will cause the testicles to reduce sperm production. This is common in professional athletes who overindulge in the therapeutic benefits of a hot tub.

In vitro fertilization is an option if it is determined that the problem lies in egg production. Fertility drugs may be administered to increase the production and release of egg cells during ovulation. If sperm count is chronically too low to even attempt to manually collect sperm cells, a sperm donor is an option.

Name: _____   Lab Section: _____

IIIIIIII Review Questions

1. Which structure carries urine from the kidney to the bladder?

_____

_____

2. Which structure in males is used for both urine and semen elimination?

_____

_____

3. The union of egg and sperm creates what?

_____

_____

4. After the completion of meiosis I, are the cells haploid or diploid?

_____

_____

5. What is the site of fetal development?

_____

_____

6. What is the normal pH range for urine?

_____

_____

7. At what point does fetal development begin?

_____

_____

8. Where in the kidney is urine formed?

_____

_____

Name: _____   Lab Section: _____

9. Label the diagram of the urinary system.

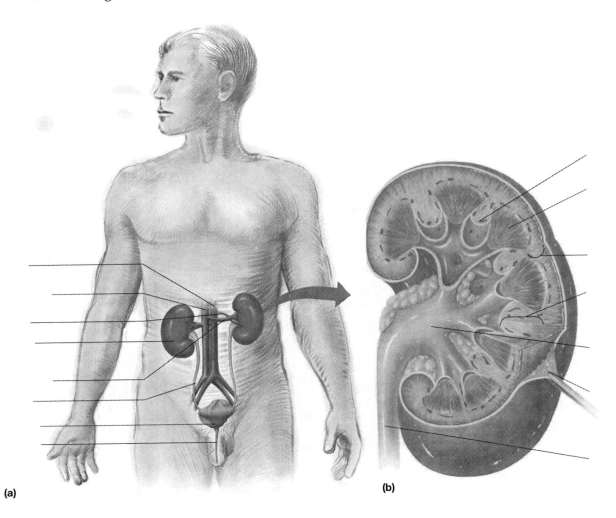

(a)

(b)

Name: _____ Lab Section: _____

10. Label the diagrams of the reproductive systems.

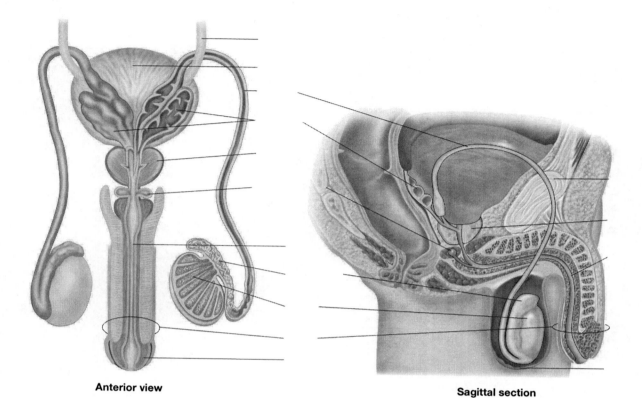

Anterior view   Sagittal section

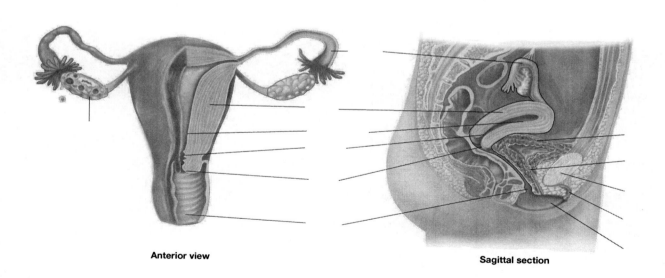

Anterior view   Sagittal section

Genetics and Heredity

- To become familiar with the principles of heredity
- To define and distinguish between DNA, chromosome, gene, allele, phenotype, and genotype
- To study genetic crosses and human traits

- PTC paper
- karyotype kits

SAFETY ALERT!

When working with human subjects, proceed humanely and respectfully. Women who are, or may be, pregnant should not perform the PTC taste test. After use, dispose of the PTC paper in a bio-hazard bag. Do not share PTC papers. Use a new one for every person.

Introduction

Genetics is the study of the patterns and processes of **heredity**—that is, the study of how traits are passed from one generation to the next. Genetic information is ultimately stored in the **DNA**, which is organized into **chromosomes**. There are 46 chromosomes in human cells except for eggs and sperm, which have 23. Exhibiting 46 chromosomes is known as the "**diploid**" condition. Eggs and sperm have the "**haploid**" number of 23.

The diploid condition of 46 chromosomes is actually the result of the pairing of the 23 chromosomes from the egg and sperm at fertilization. The chromosomes are organized into 23 **homologous** pairs, each pair carrying genes for the same traits. There are 22 pairs of **autosomes** and one pair of **sex chromosomes**. Autosomes are numbered 1 through 22, and the sex chromosomes are designated as **X** or **Y**. Females have two X chromosomes, and males have an X and a Y.

Chromosomes are further organized into discrete units called **genes**, small pieces of DNA that code for all of the traits in an organism. Genes on the paired homologous chromosomes that code for the same trait are called **alleles**.

Alleles can be either **dominant** or **recessive**. Dominant alleles will "dominate" over recessive alleles when they are paired to express traits. The combination of genes that you possess is known as your **genotype**. The traits that they express are called the **phenotype**.

In reality, genetics is much more complex. We focus on several aspects of inheritance that allow for an understanding of basic genetic principles.

In this exercise, you will study chromosomal and genetic inheritance. This is accomplished though the observation of phenotypes and determination of genotypes by employing genetic crosses.

procedure

Chromosomal Inheritance

Chromosomal inheritance clearly determines gender. However, it also accounts for other, sometimes abnormal traits. During meiosis, chromosomes sometimes fail to separate properly, a phenomenon known as **nondisjunction**. In this case, sperm and egg cells may end up with more or less than 23 chromosomes.

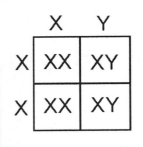

Figure 16-1 represents a cross between a female and male, both of whom have produced normal eggs and sperm. A **Punnet Square** is employed to study crosses. There are four squares that represent the possible chromosomal outcomes of the offspring. Because males and females both produce paired sex chromosomes, we have to account for the possibility of males contributing either an X or a Y chromosome and females contributing one of two X chromosomes.

Figure 16-1 Chromosomal inheritance determining gender.

 *Notice that the probability of a male or female offspring is 50%.

Nondisjunction of sex chromosomes can lead to several abnormalities. Nondisjunction in females can produce egg cells with either zero or two X chromosomes. This is denoted as **O** and **XX**. Males can produce **O**, **XY**, or **XX** sperm cells during nondisjunction.

It is standard convention to label female gametes on the left and male gametes on the top of a Punnet Square.

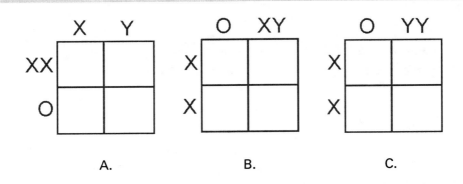

A.   B.   C.

Figure 16-2 Abnormal inheritance of sex chromosomes. A. Normal male. B. Normal female. C. Normal female.

1 In the Punnet Squares provided in **Figure 16-2**, perform crosses using the combinations when females experience nondisjunction and then when males experience nondisjunction.

2 Match the abnormal offspring to the syndromes listed here:

XO—Turner syndrome: Considered female. The lack of the second X chromosome leads to nonfunctional ovaries. This is in turn prevents the onset of puberty and development of secondary sex characteristics such as enlarged breasts and wide hips. Affected individuals will usually exhibit below average intelligence and a somewhat short, stalky, masculine build.

XXX—Poly-X syndrome: Considered female. Most develop normally and are fertile but may have slight learning disabilities.

XXY—Klinefelters syndrome: This is the most common chromosomal abnormality. As soon as a Y is present, offspring are considered male. These males are usually normal in appearance, but many have learning disabilities. The testes are usually underdeveloped, leading to sterility and poor muscle development.

XYY—Jacob syndrome: These males tend to be large and tall due to an overproduction of testosterone. They also exhibit somewhat below-average intelligence.

Karyotyping

A karyotype is a display of all the paired chromosomes in a cell. In this exercise you will examine photographs of karyotypes using the supplied kits.

1. Examine the karyotypes you have been given.

2. Cut out the individual chromosomes and pair up the homologs.

3. Determine if the individual is male or female and if they have any chrosomal abnormalities. Record the results:

_____

Genetic Inheritance

Chromosomes carry suites of genes coding for many traits. However, crosses can be done following individual genes instead of the entire chromosome. This is known as genetic inheritance. It allows for a study of one trait per cross, or a **monohybrid** cross.

Recall that you will receive one gene for each trait from both parents. Therefore, you have two genes coding for each trait. Some genes are dominant, whereas others are recessive. If you have two dominant genes, your genotype is **homozygous dominant**. Possession of two recessive genes gives you a genotype of **homozygous recessive**, and if you have one dominant and one recessive gene, you are **heterozygous**. In a genetic cross, capital letters are used to designate dominant genes, and lowercase letters are employed for recessive genes.

In a hypothetical scenario using the letter **A**, the homozygous dominant, homozygous recessive, and heterozygous conditions would be written as **AA**, **aa**, and **Aa**. The genotypes **AA** and **Aa** will express the dominant phenotype for a trait, and the genotype **aa** will express the recessive. In order for the recessive version of a trait to be expressed, two recessive genes need to be present.

In this exercise, you will observe your lab partners phenotypes for selected human traits. You will then attempt to determine the genotypes coding for the traits.

1. In pairs or groups of three or four, study each other's phenotypes for the following traits, and record the results in **TABLE 16-1**. Diagrams of select genetic traits are found in Appendix 3.

PTC taste: Place the PTC paper on your tongue for a few seconds. The ability to taste this chemical is dominant.

Attached earlobes: Observe your partners' earlobes. Attached earlobes is dominant over unattached.

Freckles: Having freckled skin is the dominant phenotype.

Tongue roll: As you extend your tongue from your mouth try to curl, or "roll" it. The ability to roll the tongue is the dominant trait.

Widows' peak: A hairline that forms a widows' peak is dominant.

Bent little finger: Place your hand with the fingers together and palm facing upward on a table. If the tip of the little finger bends towards the fourth finger you have the dominant trait.

Middigital hair: Having hair on the backside of the fingers between the knuckles is a dominant trait.

Hitchhikers thumb: Having a thumb that overextends is recessive.

Dimples: The presence of dimples is a dominant trait.

Blaze: The presence of a streak of hair that is different color than the rest of the hair on your head is a dominant trait.

Interlocking digits: Clasp together your hands by interlocking your fingers. If the left thumb is above the right thumb you possess a dominant trait.

Eye pigmentation: If you have pigment on the front of the iris of your eye, you have the dominant trait of dark eyes: black, brown, green, or hazel. You have the recessive trait of lack of pigment if your eyes are blue or gray.

2 After you record all of the phenotypes, attempt to determine genotypes and record them in Table 16-1.

TABLE 16-1	**Phenotypes and Possible Genotypes**	
Trait	**Phenotype**	**Possible Genotypes**
PTC taste (P, p)		
Attached earlobes (A, a)		
Freckles (F, f)		
Tongue Roll (T, t)		
Widows' Peak (W, w)		
Bent little finger (L, l)		
Middigital hair (M, m)		
Hitchhiker's thumb (H, h)		
Dimples (D, d)		
Blaze (B, b)		
Interlocking Digits (I, i)		
Eye Pigmentation (E, e)		

3 If you cannot definitively identify the genotype, check the phenotypes of your parents and grandparents if possible.

procedure *continued*

Sex-Linked Genes

Genes that are carried on the **X** and **Y** chromosomes are referred to as "sex-linked." One such gene found on the **X** chromosome codes for **normal vision** or **color blindness**. The gene that codes for color blindness is recessive. Because males only have one X chromosome, one recessive gene will give the color blind phenotype. Females who are heterozygous (one normal gene and one color blind gene) will have normal vision but will be considered a **"carrier"** of colorblindness.

Complete the crosses in **Figure 16-3** to determine whether offspring will have normal vision or color blindness. The normal vision gene is **N** and the color blind gene is **n**.

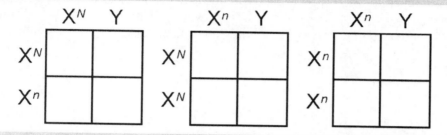

Figure 16-3 Genetic crosses for color blindness and normal vision.

Multiple Alleles and Co-Dominance

Traits such as blood type are controlled by multiple alleles. In this case, although you only get two (one from the egg and one from the sperm), there are three to choose from: A, B, and O, A and B are co-dominant, and O is recessive. Possible genotypes and phenotypes are listed here:

Genotypes	Phenotypes
AA	Type A
AO	Type A
AB	Type AB
BB	Type B
BO	Type B
OO	Type O

1 Draw a few Punnet squares to cross some of the genotypes.

2 If you know your blood type, determine your possible genotypes.

Downs Syndrome

Nondisjunction can affect any of the autosomes as well as the sex chromosomes. A condition known as **Downs Syndrome** is the result of the nondisjunction of autosomal pair number 21. In this condition, the offspring receives a third number 21 chromosome. Therefore, the condition is technically referred to as trisomy 21. This nondisjunction event usually occurs in egg cells with the probability increasing as women age. The chance of conceiving a child with Downs Syndrome steadily increases from 1 out of 3,000 births at age 30 to 1 out of 9 births at the age of 48 years.

Affected individuals all share a suite of characteristic traits: slight to severe mental retardation, wide and flattened faces, large tongue, short and stocky build, and problems associated with the heart. They usually only live until about 30 years of age and rarely reach sexual maturity.

The condition was once referred to as Mongolism because of the slight similarity in appearance to the population of Asians known as Mongols. Clearly, we realize now that this is insensitive as well as politically and socially incorrect to associate an entire population of people with a genetic disorder.

Name: _____   Lab Section: _____

IIIIIII Review Questions

1. What term is used to describe your physical appearance?

_____

_____

2. How many chromosomes are in human egg and sperm cells?

_____

_____

3. How many recessive genes are found in an individual with a heterozygous genotype?

_____

_____

4. Studying the pattern of inheritance of one trait involves what type of genetic cross?

_____

_____

5. An offspring with Klinefelters syndrome is the result of what process that can take place during meiosis in one of the parents?

_____

_____

6. Define and distinguish between the following pairs of terms: haploid/diploid, chromosome/gene, and homozygous/heterozygous.

_____

_____

_____

_____

_____

_____

_____

_____

_____

_____

LAB REPORT

7. Genetic Problems. Using genetic crosses solve the following:
   A. In a genetic cross between a male who has a heterozygous phenotype and a female with a homozygous recessive phenotype, what percentage of the offspring will possess the dominant trait?

   _____

   _____

   B. Can a male and female both exhibiting the recessive phenotype conceive a child with the dominant phenotype?

   _____

   _____

   C. In a cross between a heterozygous female and a heterozygous male, what will be the offspring phenotypic and genotypic ratios?

   _____

   _____

   D. How many of the male offspring will be color blind if a color-blind female mates with a male with normal vision? How many of the female offspring will be color blind?

   _____

   _____

   E. Can a man and a woman both with type A blood conceive a child together who has type O blood? Prove it through a genetic cross.

   _____

   _____

   _____

Human Evolution: Taxonomy and Systematics

- To define and distinguish between evolution, natural selection, taxonomy, systematics, and phylogeny.
- To survey evidence supporting evolution and simulate natural selection.
- To learn the hierarchical levels of classification and employ the methods of systematic analysis as they apply to human evolution.

- colored beads or jellybeans
- colored construction paper
- microscopes
- slides and models of embryological development
- skeletons from several vertebrates (human, cat, frog, bird, etc.)

SAFETY ALERT!

Handle the skeletal material, especially human remains, with care and respect. Carry the microscopes with two hands. Slides are made of glass and are sharp even if not broken. In the event that a slide is broken, seek guidance from the instructor concerning clean-up procedures.

IIIIII Introduction

Biological, or organic, **evolution** is the change in organisms over time. Evolution is now accepted as a theory based upon many lines of evidence. Those lines of evidence include data from studies of **comparative anatomy, DNA, embryology, the fossil record, geography**, and **geology**. The theory of evolution is now universally embraced in the "scientific" community. As part of the living world, humans are also products of evolutionary descent.

Natural selection is widely accepted as the primary mechanism for evolutionary change. Through natural selection, the "best" traits or characteristics found in organisms are "selected" by environmental constraints. The best traits are those that allow for survival and ultimately successful reproduction. Therefore, these traits will be passed on to future generations.

Taxonomy is the science of naming and classifying organisms. **Systematics** is the study of the evolutionary relationships among organisms. Ideally, a system of classification will reflect evolutionary descent. Therefore, taxonomy and systematics are often unified into one science.

In this exercise, you will survey evidence that supports the theory of evolution. You will also engage in a natural selection experiment, learn the hierarchical system of biological classification, and perform a systematic analysis.

procedure

Evidence Supporting Evolution

Comparative anatomy was one of the first lines of evidence that led the march toward an evolutionary view of life. This is easy to appreciate when you view the skeletons of the several different vertebrates shown in **Figure 17-1**. Why should organisms as diverse as humans and bats have the same bones in the same anatomical regions when each uses these bones in strikingly different ways? The logical answer is that they share a **common ancestor** and the similarities represent "variations on a common theme." These variations that are derived from a common ancestor are known as **homologous** structures.

The study of the development of organisms, formally known as embryology, quickly followed comparative anatomy in revealing evolutionary patterns.

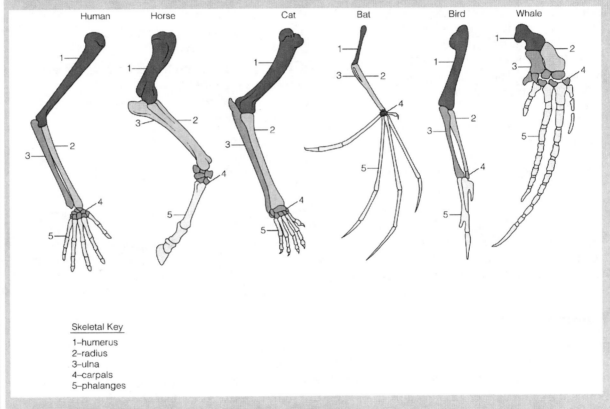

Skeletal Key

1–humerus
2–radius
3–ulna
4–carpals
5–phalanges

Figure 17-1 Homologous skeletal structures among vertebrates.

1 Observe Figure 17-1 to study the similarities and differences in the limbs of the selected vertebrates.

2 Study all available skeletal material. Study the same few bones in all organisms. In **TABLE 17-1** list a few of the bones of the organisms you studied, their location, and function.

procedure *continued*

TABLE 17.1 Comparative Skeletal Analysis.

Animal	Bone	Size	Shape	Location	Function

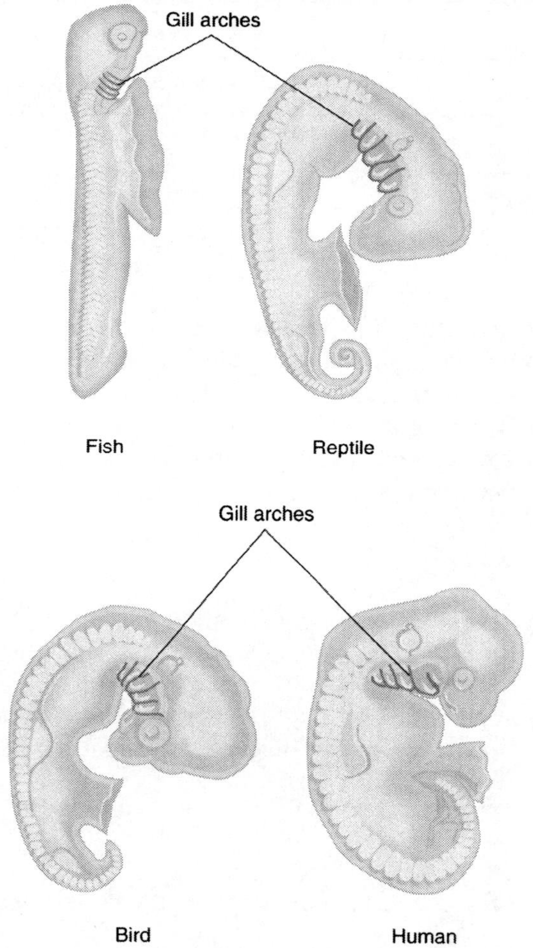

Figure 17-2 Comparisons of early embryos of several vertebrates.

3 Study **Figure 17-2** and the models and/or slides of the embryological stages of selected vertebrates. Record the similarities and differences:

____

____

_____

_____

_____

_____

_____

_____

_____

_____

_____

_____

Natural Selection

In order for natural selection to occur, three criteria must be met: (1) In a population of organisms, there must be **variability** among traits (e.g., dark vs. light skin). (2) The variability must be **inherited**. (3) Some of the variants must allow for better **reproductive success.**

In this exercise, you will simulate the process of natural selection. Jellybeans or colored beads will represent individuals in a population, and the various colors of construction paper will represent different environments. You and your lab partners will act as predators preying on the jellybeans or beads.

1 On your lab table, lay out several colors of construction paper.

2 On each colored paper, spread out 20 to 30 jellybeans or colored beads of mixed colors. Record how many of each color you start with on every paper.

_____

3 Sit down in front of one of the colored papers populated with jellybeans or beads.

4 Close your eyes for 5 to 10 seconds. Open your eyes, and immediately pick up the first 5 beans or beads that you see.

5 Repeat this two or three times.

6 For every bean or bead that is left on the paper, add two more of the same color. This represents reproductive success and the inheritance of the parent color.

7 Examine the numbers and colors of the beans or beads you now have left. Do any colors now dominate? Record the results

_____

8 Repeat steps 2 through 7 for each colored paper.

9 If natural selection has occurred, the different papers will have different colored populations of beans or beads. Record which colors of beans or beads were left on each paper.

_____

Taxonomy

Humans have a basic need to classify and categorize everything. This allows for easy storage and retrieval of information. If I told you that I live in a brick, Cape-Cod-style house, you would have a general idea of what my house looked like because all Cape-Cod houses share certain traits. This type of recognition is useful for identification and comparisons.

Taxonomy is the science of naming and classifying organisms based on shared traits. Carl Linneaus, hailed as the father of taxonomy, was one of the first to understand the importance of such work and set about organizing and categorizing organisms. He detailed his scheme of classification in his 1735 book called the _Systema Naturae_ (_System of Nature_).

Traditional taxonomy recognizes seven hierarchal categories. These range from the most inclusive kingdoms to the least inclusive level of species. Specific groups of organisms classified into these categories are known as **taxa** (singular **taxon**). There are no definitions for the "man-made" levels of kingdom through genus. The least inclusive category, the **species**, however, has always been recognized as a "real" and natural unit. The most widely accepted definition for a species is **a group of interbreeding organisms that are reproductively isolated from any other such group.**

Here is an example comparing the classification of a human, a cat, and a crow.

	Human	Cat	Crow
Kingdom	Animals	Animals	Animals
Phylum	Vertebrates	Vertebrates	Vertebrates
Class	Mammals	Mammals	Birds
Order	Primates	Carnivores	Passerines
Family	Hominids	Felids	Corvids
Genus	_Homo_	_Felis_	_Corvus_
Species	_sapiens_	_domesticus_	_americanus_

According to this scheme of classification, humans, cats, and crows share traits that include all of them in the taxa animals and vertebrates. Humans and cats, however, share traits that make both of them mammals to the exclusion of birds.

The scientific naming of organisms is also attributed to Linneaus. He devised a binomial (two-name) system to identify all organisms. The names include the genus and species level of categories. For example, humans are technically known as _Homo sapiens_. The genus level is capitalized, and the species level is in lower case. Both are either italicized or underlined.

Systematic Analysis

Systematics is the science of organizing and analyzing traits in order to build evolutionary trees called **phylogenies**. These trees exhibit a branching pattern and reflect common ancestry. After data are collected, they are put into a matrix that includes all of the taxa and traits in the study. **TABLE 17-2** is a matrix using the species in the example above: humans, cats, and crows. From this matrix, we can build a phylogeny **Figure 17-3**, a branching pattern of evolutionary history.

Ideally, all formal taxa should represent true, historical groups. These are known as **monophyletic** groups or **clades**. A monophyletic group includes a common ancestor and all of its descendants can be identified by exclusive shared traits. In our example, animals, birds, and mammals are all monophyletic groups: vertebrates have backbones, mammals have hair and mammary glands, and birds have feathers.

TABLE 17.2	Data Matrix for Human, Cat and Crow Traits				
	Animal	Backbone	Hair	Mammary Glands	Feathers
Human	x	x	x	x	
Cat	x	x	x	x	
Crow	x	x			x
Earthworm	x				

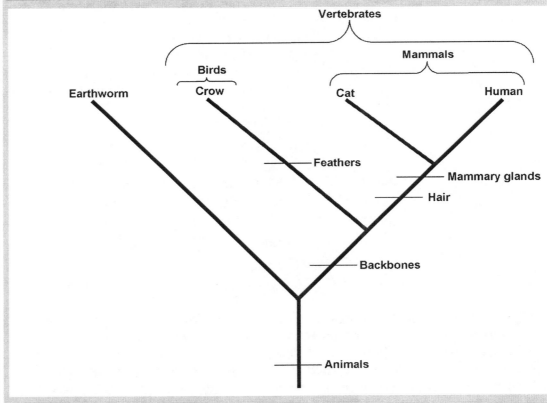

Figure 17-3 Phylogenetic relationships between humans, cats, and crows.

TABLE 17-3	Matrix of Vertebrate Traits													
	animals	limbs	hair	bipedal	reduced hair	broad incisors	large brain	opposable thumb	more erect posture	large genitals	feathers	loss of tail	mammary glands	backbones	deuterostome
Monkey	x	x	x					x					x	x	x
Human	x	x	x	x	x	x	x	x	x	x		x	x	x	x
Trout	x													x	x
Gorilla	x	x	x		x		x	x	x			x	x	x	x
Crow	x	x									x			x	x
Chimpanzee	x	x	x		x	x	x	x	x	x		x	x	x	x
Earthworm	x														
Orangutan	x	x	x				x	x	x			x	x	x	x
Cat	x	x	x											x	x
Australopithecus	x	x	x	x	x	x	x	x	x	x		x	x	x	x
Starfish	x														x

1. Using the matrix in Table 17-3, build a phylogeny of the primate and other vertebrate species.

2. Identify all monophyletic groups.

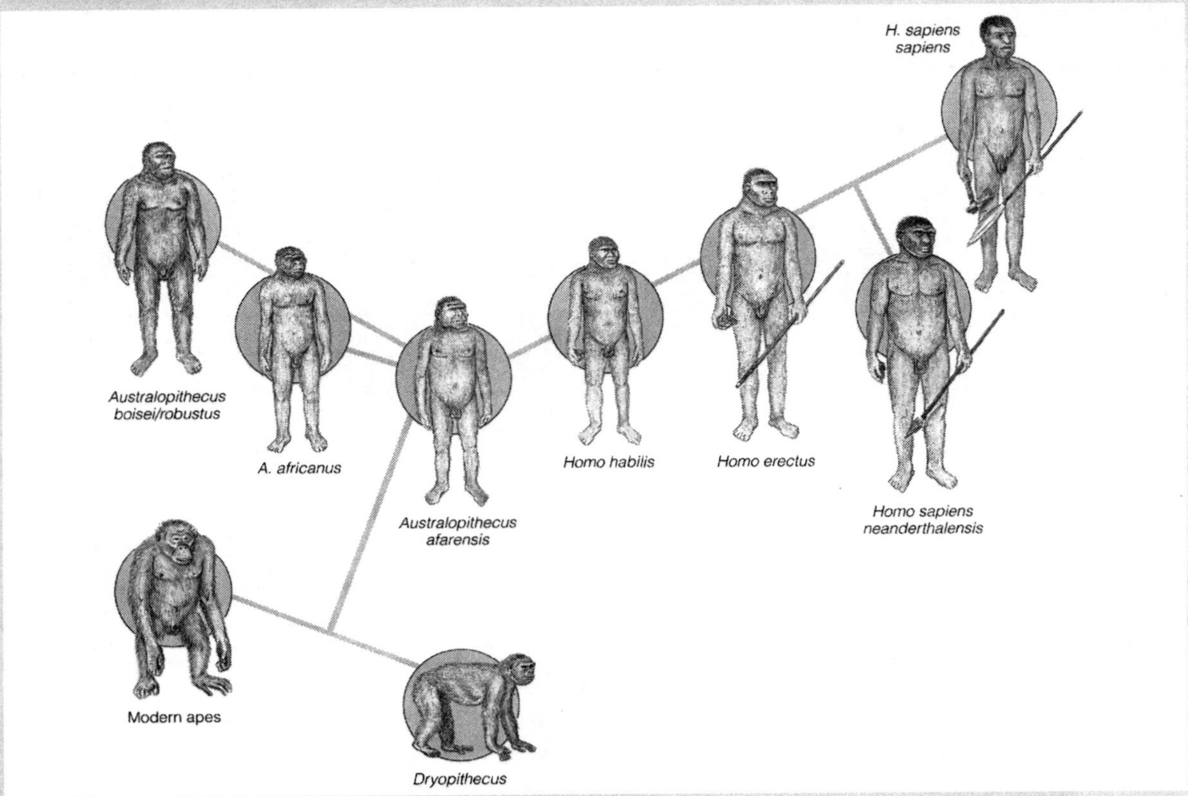

Figure 17-4 Evolution of *Homo sapiens*.

Human Evolution

Humans belong to the monophyletic family **Hominidae**, or the **Hominids**. Hominids are primates that are bipedal, that is, they walk upright on two feet. Humans are the only surviving species of hominids. All others are extinct. There exists, however, a rich fossil record that has been used to develop hypotheses of hominid phylogeny.

Figure 17-4 depicts one of the hypotheses of hominid and human evolution. **Figure 17-5** represents a phylogeny based upon skull anatomy and includes a timeline. Both of these evolutionary trees support the hypothesis that the genus *Australopithecus* (southern ape-man) is the most recent common ancestor of all hominids in the genus *Homo*.

1. Describe the anatomical trends in the evolution from apes to hominids.

2. Describe the anatomical trends in the evolution of hominids from the genus *Australopithecus* to the genus *Homo*.

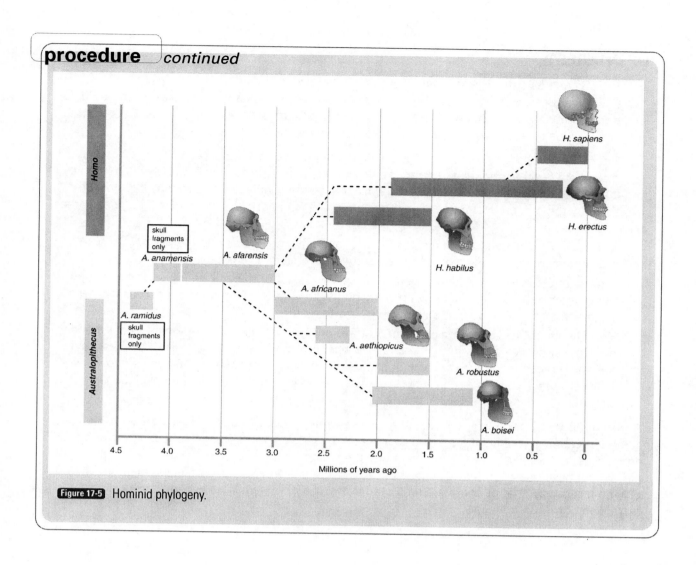

Figure 17-5 Hominid phylogeny.

Evolution Versus Intelligent Design

Have you ever heard someone say "but evolution is just a theory not a fact?" Clearly, some do not understand how the term theory is applied in science. A theory is supported by data, or "facts," from many lines of evidence and has explanatory power. Evolution meets these criteria.

Intelligent Design has been proposed as an alternative to evolution to explain the diversity and complexity of life on earth. The basic argument is that living systems are so complex with many, varied intricate relationships, especially at the cellular and molecular level, that they can only be explained by the presence of an intelligent designer. The concept of Intelligent Design faces some huge obstacles.

- Intelligent Design itself is not a scientific endeavor. It is not structured to collect data in an attempt to formulate and test hypotheses about diversity and complexity.

- Intelligent Design, by invoking supernatural causes, calls for the end of any scientific inquiry about the origins and diversity of life. Why bother to engage in any further rigorous studies if supernatural forces that are "unknowable" are at work?

- Proponents of Intelligent Design claim that evolution cannot explain everything. Most theories do not explain everything in their respective sciences. Science is progressive, however, and we know much more about evolution and its mechanisms than we did 100 years ago. There is every reason to believe that further studies will bring greater insights into evolutionary theory.

A truly Intelligent Designer, much like an engineer would not have "built" organisms with the same structures in roughly the same places that are used very differently. For instance, why would an Intelligent Designer build a human arm and the wing of a bat from the same bones in the same places, but have one used for holding and grasping and the other for flying. An engineer would only do it this way if he or she were restricted in the use of materials. In evolution we call this restriction **common ancestry**.

Name: _____ Lab Section: _____

||||||| Review Questions

1. Primates that walk on two feet are called what?

   _____

   _____

2. List three lines of evidence for the occurrence of evolution.

   _____

   _____

   _____

3. What is the primary mechanism for evolutionary change?

   _____

   _____

4. Which traits unite all mammals to the exclusion of birds?

   _____

   _____

5. What is a phylogeny?

   _____

   _____

   _____

   _____

6. Define and distinguish between the following pairs of terms: taxonomy/systematics and clade/taxon.

   _____

   _____

   _____

   _____

   _____

   _____

Name: _____ Lab Section: _____

7. What are the three criteria needed for natural selection to occur?

_____

_____

_____

8. Using the matrix, build a phylogeny.

Taxa	Traits							
	1	2	3	4	5	6	7	8
A				x	x	x		x
B								x
C	x	x	x	x		x		x
D	x	x	x	x		x	x	x
E		x		x		x		x

9. Identify the bones on the vertebrate limbs.

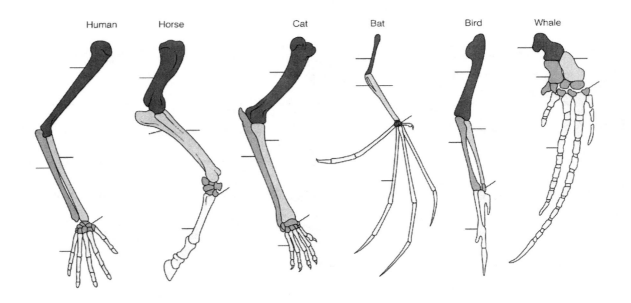

Human   Horse   Cat   Bat   Bird   Whale

Ecology

objectives

- To define and distinguish between ecology, population, community, and ecosystem
- To engage in a population estimate
- To study the effects of acid rain
- To meausure photosynthetic activity

materials

- meal worms
- aquarium
- saw dust
- calculators
- HCl acid
- pH paper
- hay infusion
- microscopes
- microscope slides and cover slips
- Paramecium culture
- Amoeba culture
- eyedroppers
- fresh elodea leaves
- test tubes
- rubber stoppers
- glass tubing
- 150 ml beakers
- lamps
- 2–3% sodium bicarbonate (baking soda) solution

⚠ SAFETY ALERT!

Carry the microscopes with two hands. Slides are made of glass and are sharp even if they are not broken. In the event that a slide is broken, seek guidance from the instructor concerning cleanup procedures. Take care not to damage or disrupt any plant life you sample on campus. Treat the mealworms humanely. Because HCl acid is caustic, exercise extreme caution when handling it.

⦚⦚⦚ Introduction

Ecology is the study of how organisms survive, or "make a living." More formally, it is the study of the interactions between organisms and their environment. The environment includes other organisms as well as the nonliving aspects such as climate, geography, and landscape.

A **population** is all of the members of a species in a defined area. A **community** is all of the members of all species in a defined area. An **ecosystem** includes the geography, landscape, and all organisms in an area. Examples of ecosystems include lakes, streams, and forests.

Ecosystems are all based on the capture and transfer of energy through food chains and webs (Figure 18-1). The ultimate source of this energy is the sun. Ecosystems are also sensitive to any changes in population and community size and changes in physical and chemical phenomena such as climate and pH.

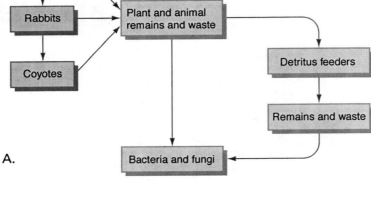

Grazer food chain

Decomposer food chain

A.

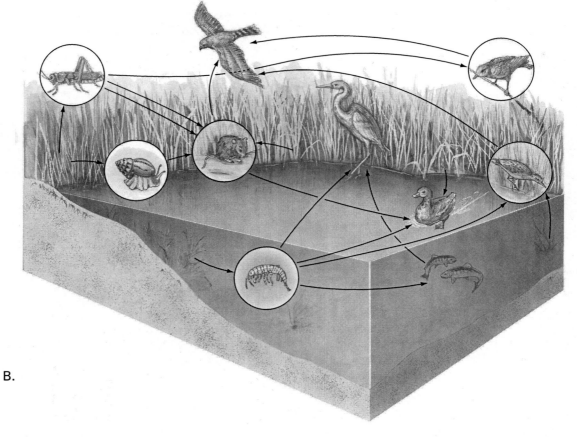

B.

Figure 18-1   A. The connections between a Grazer food chain and a decomposer food chain. B. Food chains are threads in larger food webs.

Most human cultures have learned how to manipulate their environments and no longer struggle with survival in the wild. For example, they build houses and mass produce food. Human action, however, has had a huge impact on all of the environments on the planet. Pollution and global warming are examples of the outcome of overpopulation and the accelerated use of resources such as fossil fuels.

In this exercise, you will engage in a population size estimate and study the effects of human overpopulation.

procedure

Population Estimate

The size of a population can provide very useful information about the status and health of an ecosystem. For example, fluctuations in population size can indicate changes in environmental conditions.

Several methods can be employed to arrive at an estimate of a population size. We use the term estimate because with human error and organism mobility we can never know for sure the exact population size. Some methods are clearly better than others for arriving at an accurate population estimate. The methods employed will be dependent on the organisms under investigation.

The most logical way to arrive at the size of a population would be to employ the **direct-count** method. As the term implies, this method requires that every member of the population is actually counted. This method works well with nonmobile organisms such as plants in a fairly small area. In a large area, however, even with plants, the task could become daunting. Therefore, another technique called **plot sampling** can be employed. This method calls for the division of the large study area into smaller "plots." In these smaller areas, the direct-count method is then employed.

The larger area is divided into four plots (**Figure 18-2**). In two of the plots, a direct-count method will be done. The total number of individuals counted from the two plots will then be averaged (total number divided by 2). This result will then be multiplied by four. The result will be the population estimate.

(Number from plot 1 + number from plot 2) / 2 × 4 = population estimate
= (4 + 3) / 2 × 4 = 14

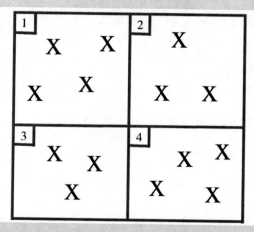

Figure 18-2 Plot sampling.

This method of estimating population assumes that all of the plots are an accurate representation of the larger area.

Plot Sampling

1 Identify an area and organism for study, for example, dandelions on campus. The area should be about 200 × 200 ft.

2 Divide the area into four 100 × 100-ft. plots.

3 Choose two of the plots to do a direct count method of population estimate.

4 Calculate the population estimate for the entire 200 × 200-ft area.

Mark and Recapture

For mobile animals, the direct-count method would clearly be very difficult to employ. Therefore, a way to estimate wild animal populations is the **mark-recapture method**. The first step in this method is to chose a study area. Individuals are then caught and marked and released back into the same area. At a later time, another attempt is made to capture individuals in the exact same defined area. With the data gathered between the first and second captures, an estimate of the population can be made. The estimate is based on simple proportion rules and is known as the **Lincoln-Peterson** method of population estimation. This method employs the following calculation:

$$P = (M \times p) / m$$

Where
P = the estimate of the total population
M = the number of individuals caught, marked, and then released
p = the total number of individuals caught in the second capture
m = the total number that were already marked in the second capture

For example, suppose that a researcher captured, marked, and released 200 gray squirrels. At a later time, 50 of the marked squirrels are captured with 100 new squirrels. The researcher could then assume that one fourth (50 of 200) of the population had been captured during the recapture procedure. Because 150 had been captured during the second study, the population size estimate would be as follows:

one fourth of P = 150 or P = 150 × 4 = 600

Using the Lincoln-Peterson Equation

P = 200 × 150 / 50 = 600

This method of population estimation can be valid and reliable only if certain basic assumptions are met:

1 Marked and unmarked individuals have the same chance of being captured; that is, marking the individuals should not increase or decrease the probability of being caught.

2 Birth and death rates between the first and second capture are small enough to be ignored.

3 After being marked and released, the individuals randomly disperse throughout the population.

4 The number of individuals moving into or out of the study area is small enough to be ignored.

Mark-and-Recapture: Using Meal Worms in an Aquarium

The instructor will have an aquarium full of a "known" number of mealworms and saw dust.

1 Reach into the aquarium and feel around for a few seconds. If you find a worm, humanely and carefully remove it.

2 Repeat the first step 10 to 20 times. You may not find a worm on every attempt.

3 **Mark** the organisms with a sharpie pen.

4 Return the organisms to the aquarium.

5 Gently mix the worms and saw dust to scatter the worms.

6 Have your lab partner then repeat the first step 10 to 20 times. This is the **recapture**.

7 Record all data from both captures, and calculate the population estimate using the Lincoln-Peterson ratio.

Population estimate _____

Population Growth Curves and the Impact of Human Population Size

The size of a population is determined by many factors. Two obvious factors are birth and death rates. Others are known as **resources**. Some resources such as nutrients and water are considered "renewable" because they cycle through ecosystems (**Figure 18-3**). Others, such as space, are nonrenewable.

Figure 18-4 graphically depicts population growth curves. The carrying capacity indicates how many individuals the environment can sustain over a long-term period.

Figure 18-5 represents human population increase over the past 2000 years. Humans are the only organisms on earth to show such a rapid increase.

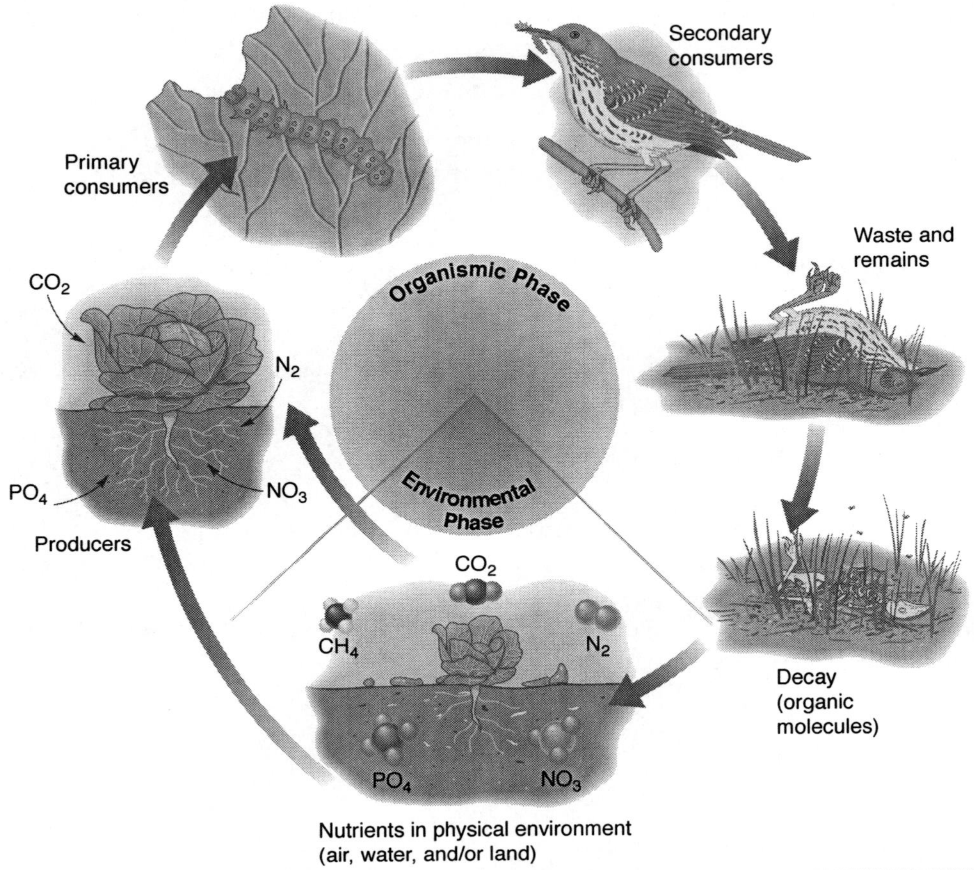

Figure 18-3 Organismic and environmental phases of nutrient cycles. Nutrients exist in organisms and their abiotic environment; they cycle back and forth between these two components of the ecosystem. CO_2 = carbon dioxide, N_2 = nitrogen, PO_4 = phosphate, NO_3 = nitrate, and CH_4 = methane.

Based on the human growth curve in Figure 18-5, hypothesize about what might account for such curve:

_____

_____

Human activity such as advanced industrial processes and burning of fossil fuels releases pollutants into the environment. One of the many outcomes of this is the creation of acid rain. This occurs because the burning of coal and other fossil fuels puts sulfur and nitrogen oxide gases into the air. These are converted to sulfuric and nitric acids in the atmosphere. Rain washes these acids back to the surface of the earth. Many organisms are adversely affected due to the shift in pH.

Measuring the Effects of pH Changes

1 Measure the pH of the pure amoeba culture, pure paramecium culture, and the hay infusion.

2 Using separate eyedroppers, prepare a slide with a cover slip for each of the cultures.

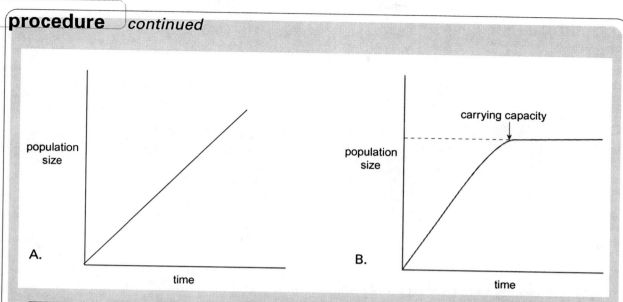

Figure 18-4 (A) Population growth with unlimited resources. (B) A population that has reached carrying capacity.

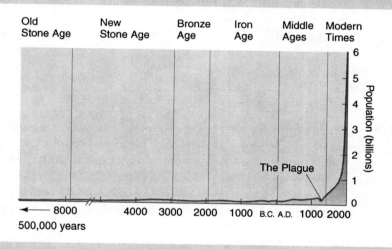

Figure 18-5 World population growth. This graph dramatically illustrates the growth curve of the world population.

3 View the slides under the microscope for organism activity (**Figure 18-6**).

4 Record your observations _____.

5 Take half of each culture and transfer each to separate clean beakers.

6 Using drops of HCl acid, lower the pH to about 4 in each of the new cultures.

7 Repeat steps 2 and 3 and record your observations. _____

_____

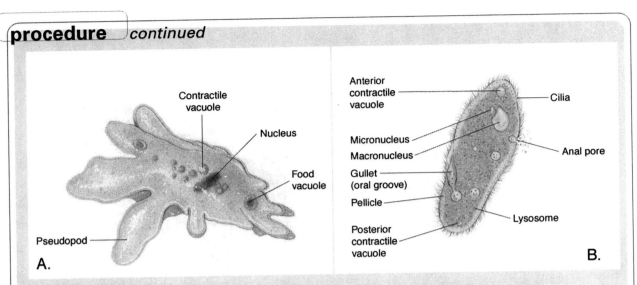

Figure 18-6 Microorganisms. (A) Amoeba. (B) Paramecium.

Energy Transformation and Human Trophic Levels

The ultimate source of energy is the sun. Green plants capture the energy from the sun in a process known as **photosynthesis**. This reaction combines carbon dioxide and water to form glucose and oxygen:

$$CO_2 + H_2O \xrightarrow{\text{sunlight}} C_6H_{12}O_6 + O_2$$

Green plants are known as **autotrophs** ("self-nourishment") because they can make their own food, glucose. Glucose is the simple sugar that most organisms use to fuel all of their cellular activities. Organisms such as animals are called **heterotrophs** because they need to eat other organisms to get their food for energy (**Figure 18-7**). Animals that eat green plants are known as **herbivores**, or primary consumers. Those that eat other animals are called **carnivores**, or secondary consumers. In a food web, some animals are **omnivores**; they eat both plants and animals (see Figure 18-1).

Biomass is the amount of mass, or weight, of organisms found in an ecosystem. In general, the closer the organisms are to the source of energy, the sun, the more biomass they will gather. Figure 18-7 illustrates this concept as the green plants far "outweigh" the top carnivores. Energy transformation from one trophic level to the next is not very efficient. Therefore, there is not enough energy to sustain a large amount of biomass at the top of the food web. This is why "big, fierce animals are rare."

Human populations that indulge in a more vegetarian diet are literally eating closer to the sun. This allows for a much more productive capture of the sun's energy. Eating a diet high in meats is a luxury that many cultures cannot afford. It costs a great deal of energy (usually grains) to feed the animals. When the animals are then eaten by humans, less energy is extracted than would have been if humans would have eaten the grains directly (**Figure 18-8**).

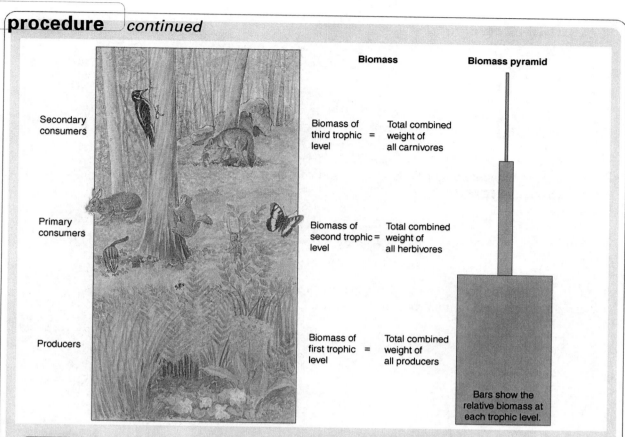

Figure 18-7 In most food chains, biomass decreases from one trophic level to the next higher one.

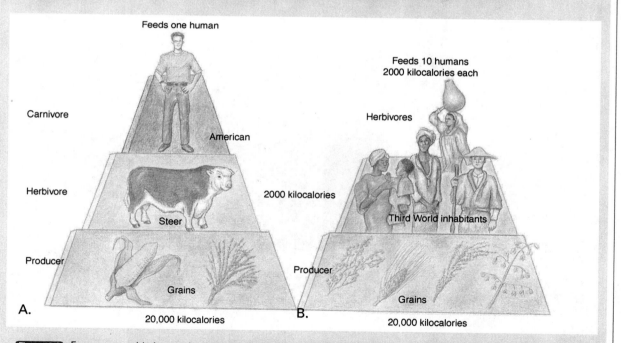

Figure 18-8 Energy pyramids in two food chains. (A) The typical meat-based diet. The 20,000 kilocalories of corn fed to cows produces only 2,000 kilocalories of meat. An adult needs only about 2,000 calories per day. (B) In a shorter food chain, 20,000 kilocalories can feed 10 people directly. This is the reason many people in developing nations subsist primarily on a vegetarian diet. Although few, if any, people eat a meat-only diet, this example does illustrate an important point: more food is available to those societies that eat lower on the food chain.

Photosynthesis

1   Place freshly cut elodea leaves into a test tube with the 2-3% sodium bicarbonate solution. This solution will provide elodea with both water and CO_2 for photosynthesis.

2   Put a glass tube into a stopper and insert the combination into the test tube. This will serve as the volumeter. **Figure 18-9** .

3   Set the test tube next to a large beaker of tap water. Place the lamp on the other side of the beaker.

4   Turn on the lamp for 10-15 minutes. As photosynthesis occurs the water level in the volumeter should rise due to the production of oxygen. Record the results _____.

_____

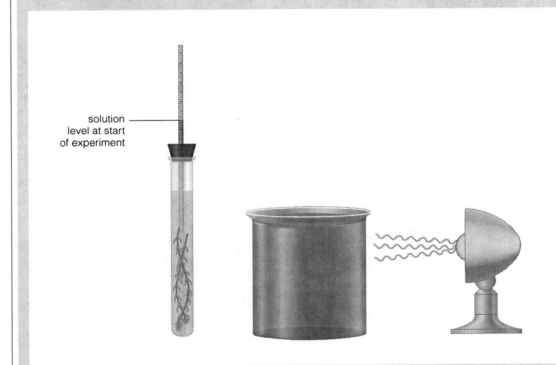

solution
level at start
of experiment

Figure 18-9 Photosynthesis volumeter setup.

OF INTEREST

Global Warming

The concept of global warming has been used to fuel political agendas much like many other environmental issues. The scenario is familiar: scientists warn of real and potential dangers to global ecosystem health, whereas capitalists tend to distort the data because of the threat to industrial profits.

Global warming is actually a natural process that occurs due to the production of waste products, mostly gases, from living systems. These green house gases include methane and carbon dioxide. They cause heat to stay in the earths' atmosphere. This helps to ensure that the earth stays warm enough to sustain living systems.

Overuse of fossils fuels such as gasoline and coal causes a dramatic increase in carbon dioxide levels. This along with deforestation, which takes away the natural "users" of carbon dioxide, the green plants, account for the phenomenon known as the greenhouse effect. The greenhouse effect is the increase in average global temperature (global warming) due to the trapping of too much heat in the atmosphere. This has been linked to the melting of the polar regions, flooding, and devastating global weather patterns. **Figure 18-10** shows the link between increased CO_2 levels and increased global temperatures.

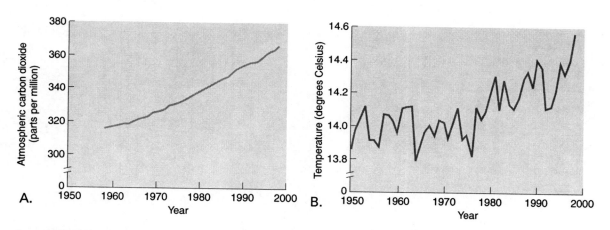

Figure 18-10 A. Carbon dioxide levels in the atmosphere have risen dramatically since 1958. B. Graph of average global temperature since 1950. Data source: GISS, BP, IEA, CDIAC, DOE, and Scripps Institute of Oceanography.

Name: _____ Lab Section: _____

IIIIII Review Questions

1 Plot sampling would be a good method to estimate population for what type of organisms?

_____

_____

2. What are the products of the reaction of photosynthesis?

_____

_____

3. What type of population estimate would you employ for grasshoppers in a field?

_____

_____

4. What is the ultimate source of energy on earth?

_____

_____

5. Which animals eat both plants and other animals?

_____

_____

6. Define and distinguish between the following pairs of terms: population/community and environment/ecosystem.

_____

_____

_____

_____

7. What is the difference between renewable and nonrenewable resources?

_____

_____

_____

_____

_____

Name: _____ Lab Section: _____

8. During a mark-and-recapture exercise you caught and marked 50 crickets in field. A few days later you attempted a recapture in the exact location. This time you caught 30 marked crickets and 40 unmarked crickets. What is the population estimate for crickets in the field?

_____

_____

9. Draw the graph of population growth with unlimited resources. Draw the graph of a population that has reached carrying capacity.

10. Describe the process that gives rise to acid rain.

_____

_____

_____

_____

_____

_____

_____

_____

_____

Anatomical Terminology, Planes, and Directions

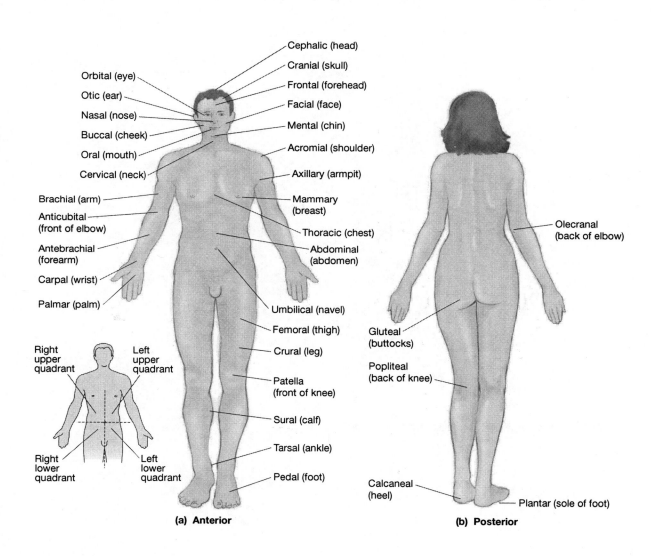

Orbital (eye)
Otic (ear)
Nasal (nose)
Buccal (cheek)
Oral (mouth)
Cervical (neck)
Brachial (arm)
Anticubital (front of elbow)
Antebrachial (forearm)
Carpal (wrist)
Palmar (palm)

Cephalic (head)
Cranial (skull)
Frontal (forehead)
Facial (face)
Mental (chin)
Acromial (shoulder)
Axillary (armpit)
Mammary (breast)
Thoracic (chest)
Abdominal (abdomen)
Umbilical (navel)
Femoral (thigh)
Crural (leg)
Patella (front of knee)
Sural (calf)
Tarsal (ankle)
Pedal (foot)

Right upper quadrant
Left upper quadrant
Right lower quadrant
Left lower quadrant

Olecranal (back of elbow)
Gluteal (buttocks)
Popliteal (back of knee)
Calcaneal (heel)
Plantar (sole of foot)

(a) Anterior

(b) Posterior

Figure App 1-1 The anatomical positions and anatomical terminology.

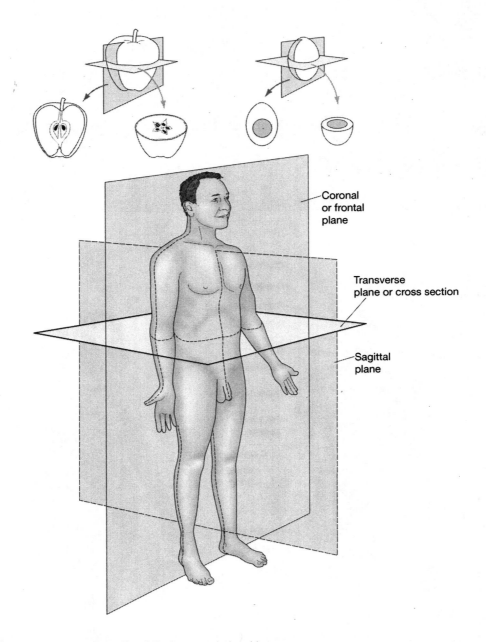

Coronal or frontal plane

Transverse plane or cross section

Sagittal plane

Figure App 1-2 Planes of section. The subject is in the anatomical position.

TABLE 1	Locational Terminology	
Term	**Definition**	**Example**
Medial	Toward the midline of the body	Your nose is medial to your eye. (Your nose is nearer to the midline than your eye.)
Lateral	Away from the midline of the body	Your ear is lateral to your eye. (Your ear is further away from your midline than your eye.)
Intermediate	Between two structures	Your eye is intermediate to your nose and your ear. (It is between your nose and ear.)
Anterior	Toward the front surface of your body	Your nose is an anterior structure. (It is on your front surface.)
Posterior	Toward the rear surface of your body	Your kidneys are posterior to your intestines. (They are behind your intestines.)
Ventral	Toward the front surface of your body	Your navel is a ventral structure. (It is located on your front surface.)
Dorsal	Toward the rear surface of your body	Your spinal cord is a dorsal structure. (It is toward your back surface.)
Superior	Above or higher than	Your head is superior to your neck. (Your head is above your neck.)
Inferior	Below or lower than	Your chest is inferior to your neck. (Your chest is below your neck.)
Superficial	Toward the surface of your body	Your skin is superficial to your muscles. (Your skin is closer to the surface than your muscles are.)
Deep	Away from the surface	Your brain is deep to your skull. (Your brain is further from the surface than your skull is.)
Ipsilateral	Same side	Your right ear is ipsilateral to your right eye.
Contralateral	Opposite side	Your left ear is contralateral to your right eye.
Proximal*	Closer to the beginning or site of attachment	Your elbow is proximal to your wrist. (Your elbow is closer to the attachment of your arm to your trunk than your wrist is.)
Distal*	Further from the beginning or site of attachment	Your colon is distal to your small intestines. (The digestive tract is a tubular structure; the colon is further from the beginning of the tube than the small intestine is.)

* Terms used in tubular structures or in the limbs.

Fetal Pig Dissection

||||||| Removal of the skin

Place the pig on its back in the tray. Pull the skin up on the midline in the abdomen. Make a small puncture with the tip of the scissors. Gently and slowly slide the scissors under the skin. While taking care to keep the tip of the scissors facing upward, follow the path of incisions shown in [Figure App 2-1]. Try slipping your fingers under the skin to gently remove it from the underlying muscle. This is sometimes easier and safer than using only scissors. **Caution:** The underlying muscles are very delicate and the skin is attached to them by way of a loose connective tissue: areolar tissue.

Extreme patience is required in order to perform this procedure without cutting too deeply and damaging the underlying muscles and internal organs.

||||||| Exposing the internal organs

Starting at the region of the genitals, carefully puncture through the skin (if not yet removed) and the body wall. Slide the scissors gently and slowly in while keeping the scissors pointed upward. Make an incision from the genitals to the mandible (chin). Splitting through the rib cage may pose a bit of difficulty. It is tempting to point the scissors downward for more force. Resist this as it can cause damage to the heart and lungs.

Using the same techniques, make lateral (to the side) incisions from the genitals, just above the hind legs, to the sides of the abdomen. Make similar lateral incisions just below the front legs.

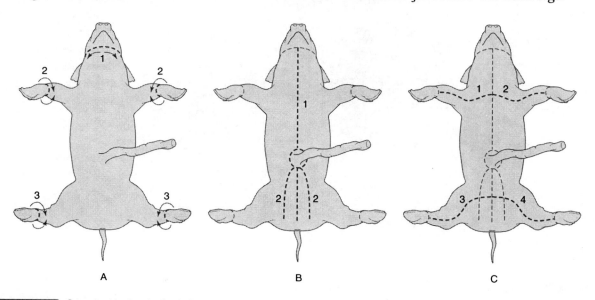

Figure App 2-1 Steps in skinning the fetal pig.

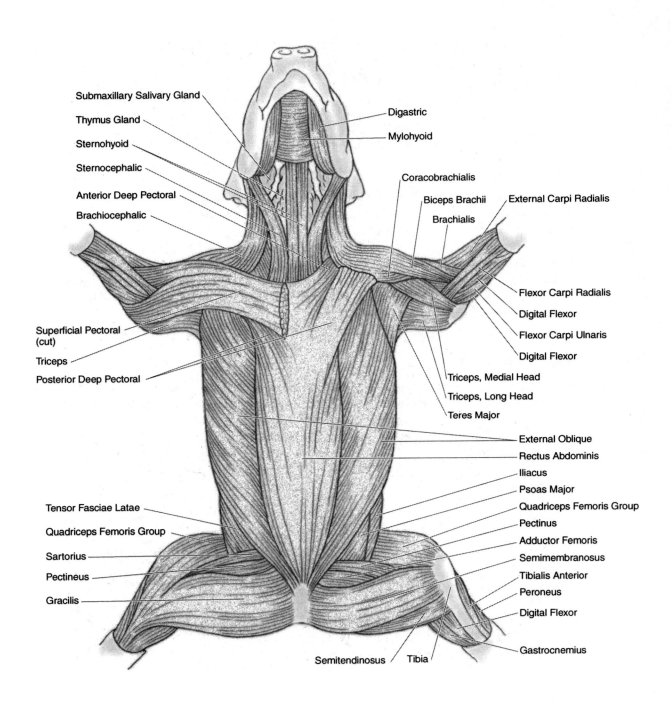

Submaxillary Salivary Gland

Thymus Gland

Sternohyoid

Sternocephalic

Anterior Deep Pectoral

Brachiocephalic

Digastric

Mylohyoid

Coracobrachialis

Biceps Brachii

Brachialis

External Carpi Radialis

Superficial Pectoral (cut)

Triceps

Posterior Deep Pectoral

Flexor Carpi Radialis

Digital Flexor

Flexor Carpi Ulnaris

Digital Flexor

Triceps, Medial Head

Triceps, Long Head

Teres Major

External Oblique

Rectus Abdominis

Iliacus

Psoas Major

Quadriceps Femoris Group

Pectinus

Adductor Femoris

Semimembranosus

Tibialis Anterior

Peroneus

Digital Flexor

Gastrocnemius

Tensor Fasciae Latae

Quadriceps Femoris Group

Sartorius

Pectineus

Gracilis

Semitendinosus

Tibia

Figure App 2-2   Ventral aspect of fetal pig: muscles of the chest and abdomen.

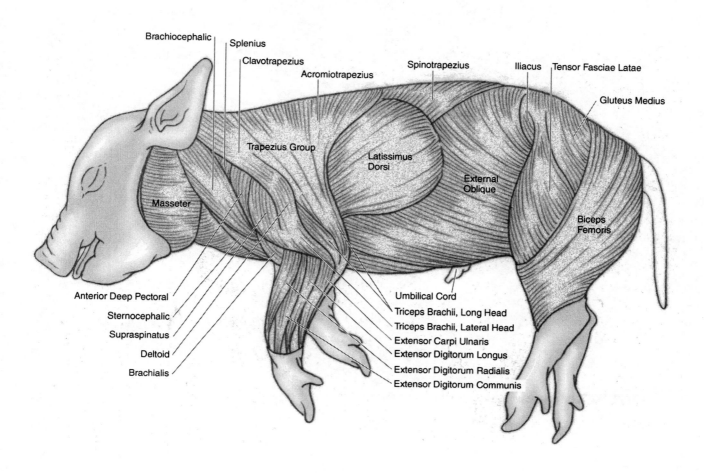

Brachiocephalic

Splenius

Clavotrapezius

Acromiotrapezius

Spinotrapezius

Iliacus

Tensor Fasciae Latae

Gluteus Medius

Trapezius Group

Latissimus Dorsi

External Oblique

Biceps Femoris

Masseter

Anterior Deep Pectoral

Sternocephalic

Supraspinatus

Deltoid

Brachialis

Umbilical Cord

Triceps Brachii, Long Head

Triceps Brachii, Lateral Head

Extensor Carpi Ulnaris

Extensor Digitorum Longus

Extensor Digitorum Radialis

Extensor Digitorum Communis

Figure App 2-3 Lateral aspect of fetal pig muscles.

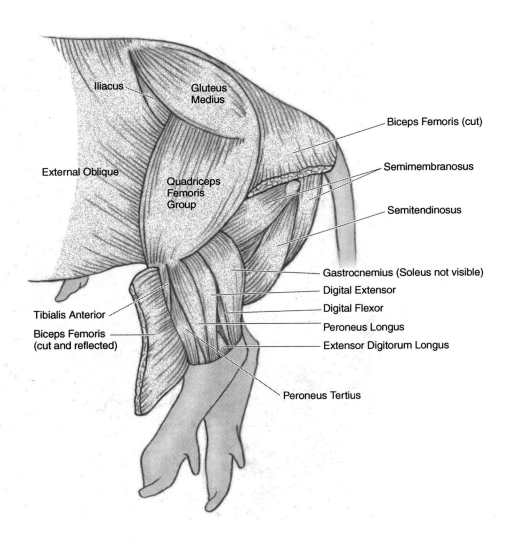

Figure App 2-4 Lateral aspect of fetal pig, left hind limb muscles.

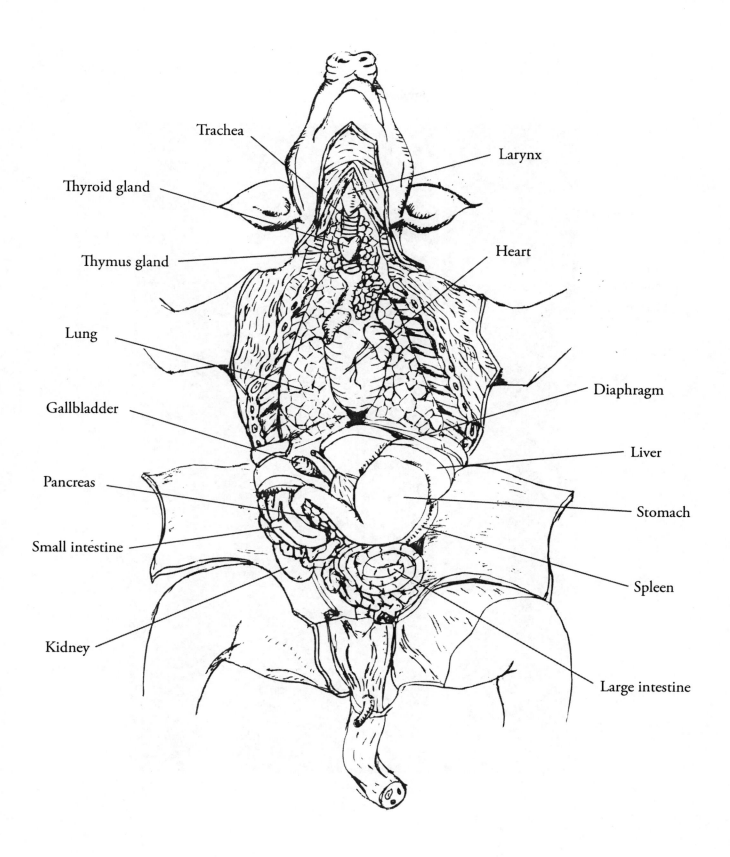

Trachea

Larynx

Thyroid gland

Thymus gland

Heart

Lung

Diaphragm

Gallbladder

Liver

Pancreas

Stomach

Small intestine

Spleen

Kidney

Large intestine

Figure App 2-5 Internal organs of the fetal pig.

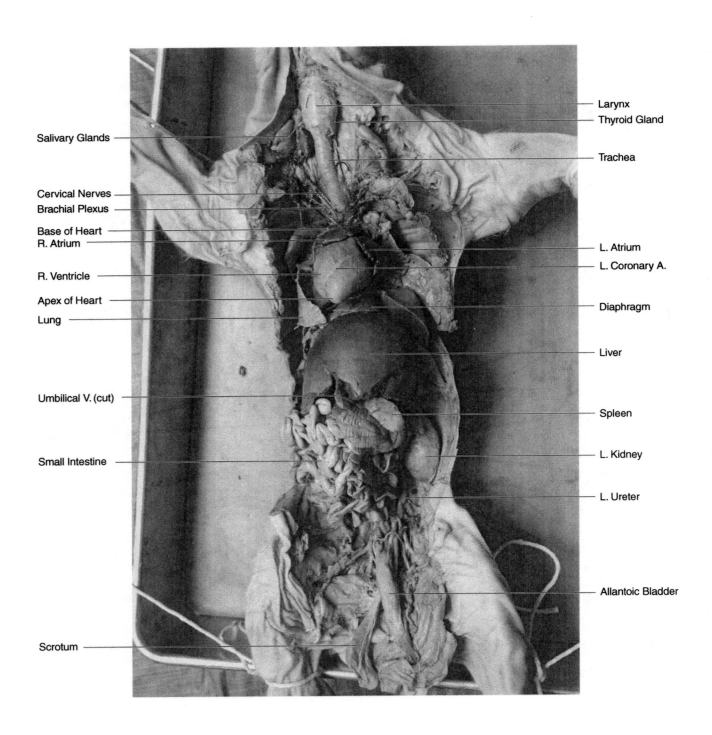

Salivary Glands

Cervical Nerves
Brachial Plexus

Base of Heart
R. Atrium

R. Ventricle

Apex of Heart

Lung

Umbilical V. (cut)

Small Intestine

Scrotum

Larynx
Thyroid Gland

Trachea

L. Atrium
L. Coronary A.

Diaphragm

Liver

Spleen

L. Kidney

L. Ureter

Allantoic Bladder

Figure App 2-6 Overview of organs of fetal pig thoracic and abdominal cavities.

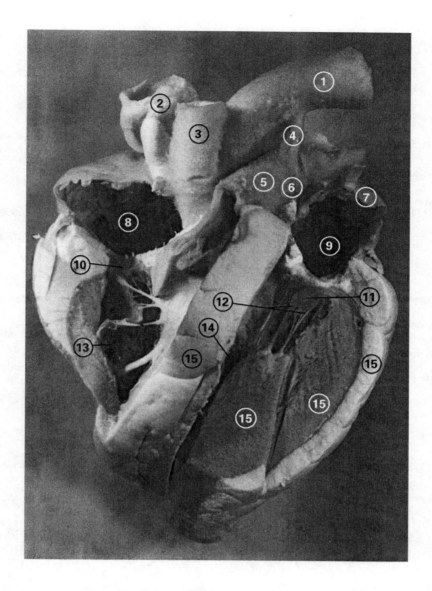

Figure App 2-7 Sheep heart, frontal section. 1. aortic arch, 2. superior vena cava., 3. brachiocephalic artery, 4. ligamentum arteriosum, 5. pulmonary trunk, 6. left pulmonary artery, 7. myocardium of left atrium, 8. right atrium, 9. left atrium, 10. tricuspid valve, 11. mitral valve, 12. chordae tendineae, 13. right ventricle, 14. left ventricle, 15. myocardium of left ventricle.

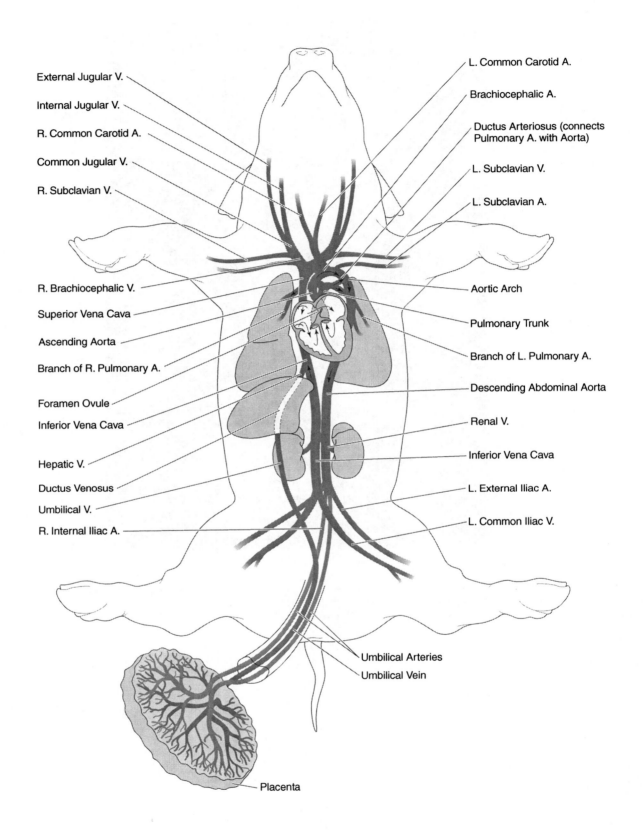

External Jugular V.

Internal Jugular V.

R. Common Carotid A.

Common Jugular V.

R. Subclavian V.

R. Brachiocephalic V.

Superior Vena Cava

Ascending Aorta

Branch of R. Pulmonary A.

Foramen Ovule

Inferior Vena Cava

Hepatic V.

Ductus Venosus

Umbilical V.

R. Internal Iliac A.

L. Common Carotid A.

Brachiocephalic A.

Ductus Arteriosus (connects Pulmonary A. with Aorta)

L. Subclavian V.

L. Subclavian A.

Aortic Arch

Pulmonary Trunk

Branch of L. Pulmonary A.

Descending Abdominal Aorta

Renal V.

Inferior Vena Cava

L. External Iliac A.

L. Common Iliac V.

Umbilical Arteries

Umbilical Vein

Placenta

Figure App 2-8 Overview of fetal pig vessels.

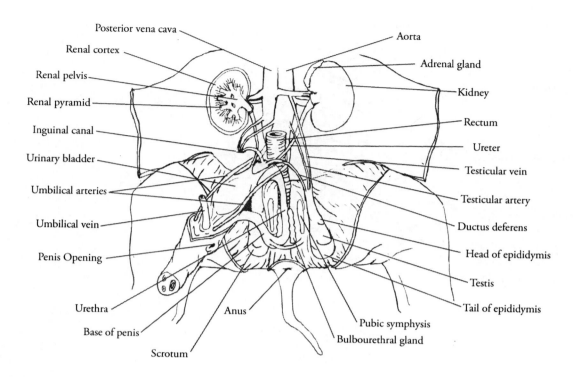

Figure App 2-9 Fetal pig male reproductive system.

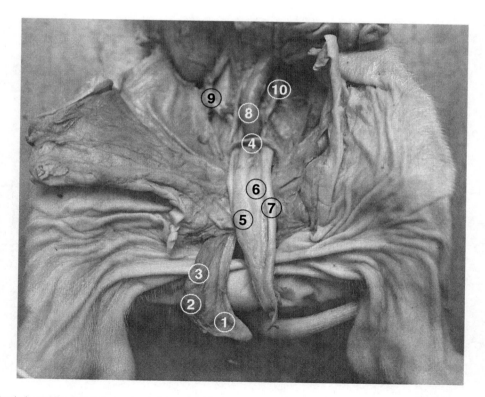

Figure App 2-10 Fetal pig male reproductive system with surrounding structures. 1. scrotal sac, 2. right testis, 3. epididymis, 4. ductus deferens, 5. penis, 6. allantonic bladder, 7. umbilical artery, 8. descending portion of colon, 9. right ureter, 10. left ureter.

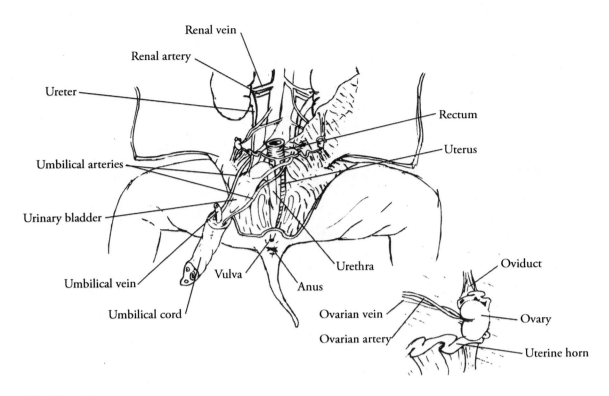

Figure App 2-11 Fetal pig female reproductive system.

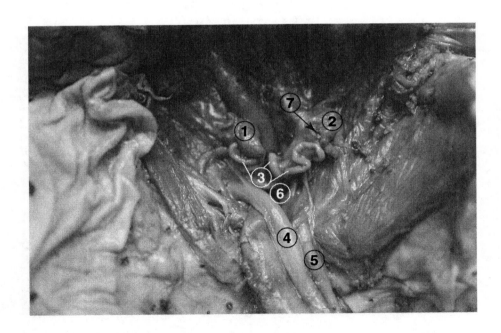

Figure App 2-12 Fetal pig female reproductive system with surrounding structures. 1. right ovary, 2. left ovary, 3. uterine horns, 4. umbilical artery, 5. allantonic bladder, 6. uterus (within), 7. fimbriae of Fallopian tube

Dominant and Recessive Genetic Traits

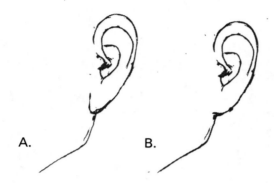

Figure App 3-1 Earlobes: A. Free—Dominant. B. Attached—Recessive.

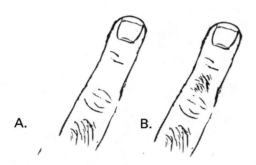

Figure App 3-2 Mid-digital Hair: A. No hair—Recessive. B. Hair—Dominant.

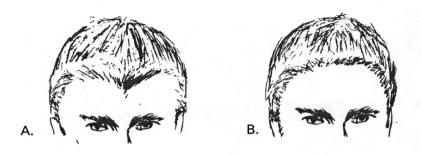

Figure App 3-3 Hairline: A. Widows' Peak—Dominant. B. No peak—Recessive.

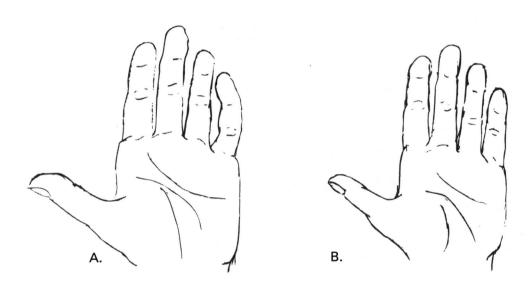

Figure App 3-4 Little Finger ("pinky"): A. Bent—Dominant. B. Straight—Recessive.

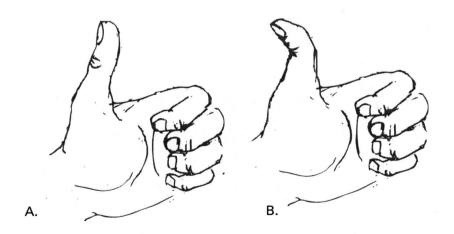

Figure App 3-5 Hitchhikers Thumb: A. Straight—Dominant. B. Overextended ("hitchhiker")—Recessive.

Photo Credits

Exercise 3

Figure 3.5a © Michael Abbey Visuals Unlimited; Figure 3.5b © Dr. John D. Cunningham/Visuals Unlimited; Figure 3.5c © Carolina Biological Supply Company/Phototake; Figure 3.5d © Dr. John D. Cunningham/Visuals Unlimited; Figure 3.5e © Michael Abbey/Photo Researchers, Inc.; Figure 3.5f © Michael Abbey/Photo Researchers, Inc.

Exercise 5

Figure 5.2 Courtesy of Anne B. Donnersberger; Figure 5.3 Courtesy of Anne B. Donnersberger; Figure 5.4 Courtesy of Anne B. Donnersberger; Figure 5.5 Courtesy of Anne B. Donnersberger; 5.6 Courtesy of Richard A. Scott, M.D.; Figure 5.7 Courtesy of Anne B. Donnersberger; Figure 5.8 Courtesy of Richard A. Scott, M.D.; Figure 5.9 Courtesy of Anne B. Donnersberger; Figure 5.11 Courtesy of Anne B. Donnersberger; Figure 5.12 Courtesy of Anne B. Donnersberger; Figure 5.13 Courtesy of Richard A. Scott, M.D.; Figure 5.14 Courtesy of Richard A. Scott, M.D.; Figure 5.15 Courtesy of Anne B. Donnersberger

Exercise 6

Figure 6.5a From Edward J. Reith and Michael H. Ross, Atlas of Descriptive Histology, 3rd edition, 1977 by Harper and Row. Used by permission of Michael H. Ross; Figure 6.5b © Dr. John D. Cunningham/Visuals Unlimited; Figure 6.5c © Dr. John D. Cunningham/Visuals Unlimited; Figure 6.5d © David M. Phillips/Visuals Unlimited

Exercise 8

Figure 8.5 Illustration by Penelope J. Nicholls; Figure 8.6 © Cabisco/Visusals Unlimited

Exercise 9

Figure 9.1 Courtesy of Richard A. Scott, M.D.

Exercise 10

Figure 10.3 © SIU/Visuals Unlimited

Exercise 11

Figure 11.2 Courtesy of Robert K. Clark, Ph.D

Exercise 13

Figure 13.3a Courtesy of Anne B. Donnersberger; Figure 13.3b Courtesy of Anne B. Donnersberger; Figure 13.3c Courtesy of Kathleen Ahearn; Figure 13.4b Illustration by Pat Oaks

Exercise 13

Figure 15.10 Courtesy of Anne B. Donnersberger; Figure 15.11 Courtesy of Anne B. Donnersberger; Figure 15.12 Courtesy of Anne B. Donnersberger

Appendices

Figure App 2.6 © Jones and Bartlett Publishers. Photographed by Kimberly Potvin; Figure App 2.7 Courtesy of Kathleen Ahearn; Figure App 2.10 © Jones and Bartlett Publishers. Photographed by Kimberly Potvin; Figure App 2.12 © Jones and Bartlett Publishers. Photographed by Kimberly Potvin

CPSIA information can be obtained at www.ICGtesting.com
Printed in the USA
LVOW092307171212

312118LV00001B/1/P